World Shipping and Port Development

World Shipping and Port Development

Edited by Tae-Woo Lee and Kevin Cullinane

First published 2005 by
PALGRAVE MACMILLAN
Houndmills, Basingstoke, Hampshire RG21 6XS and
175 Fifth Avenue, New York, N. Y. 10010
Companies and representatives throughout the world

PALGRAVE MACMILLAN is the global academic imprint of the Palgrave
Macmillan division of St. Martin's Press, LLC and of Palgrave
Macmillan Ltd. Macmillan® is a registered trademark in the United
States, United Kingdom and other countries. Palgrave is a registered
trademark in the European Union and other countries.

ISBN-13: 978–1–4039–4753–6
ISBN-10: 1–4039–4753–8

This book is printed on paper suitable for recycling and made from
fully managed and sustained forest sources.

A catalogue record for this book is available from the British Library.

Library of Congress Cataloging-in-Publication Data
World shipping and port development / edited by Tae-Woo Lee and
Kevin Cullinane.
    p. cm.
  Includes bibliographical references and index.
  ISBN 1–4039–4753–8 (cloth)
    1. Shipping. 2. Harbors. I. Lee, Tae-Woo. II. Cullinane, Kevin.

HE571.W674 2005
387–dc22   2004060786

10   9   8   7   6   5   4   3   2   1
14   13   12   11   10   09   08   07   06   05

Printed and bound in Great Britain by
Antony Rowe Ltd, Chippenham and Eastbourne

Contents

List of Figures

List of Tables

Notes on the Contributors

Dr Alfred Baird is Head of the Maritime Research Group at Napier University's Transport Research Institute (TRI) in Edinburgh. Prior to his academic career he worked for a liner shipping company, followed by a position as shipping manager in a manufacturing company. He has researched, published and taught in a number of areas in maritime transport, covering issues such as port competition and privatization, strategic management in shipping, market analysis in the shipping and ports industry, shipping cost modelling, and assessing the feasibility of new shipping services and associated port developments.

Dr Marco Benacchio was born in 1973 and has a PhD in 'Transport Economics'. He is currently case handler at the Italian Competition Authority, Directorate of Transport and Manufacturing. He is a lecturer on the course 'Maritime and Port Economics' at the University of Genoa. His main research areas are in the field of transport economics where he is co-author of more than 30 scientific publications.

Prof. Mary R. Brooks is the William A. Black Chair of Commerce at Dalhousie University, Halifax, Canada. She was Membership Secretary and Treasurer of the International Association of Maritime Economists from 1994 to 1998 and a Director of the Halifax International Airport Authority from 1995 to 2004. She currently chairs the Committee on International Trade and Transportation, Transportation Research Board, Washington DC. She is a Member of the Chartered Institute of Logistics and Transport. In 2005, she will be a Canada–US Fulbright Scholar based at George Mason University in Fairfax, VA. Dr Brooks received her undergraduate degree from McGill University (1971), her MBA from Dalhousie University (1979) and her PhD in Maritime Studies from the University of Wales in 1983.

Prof. Kevin Cullinane is Chair in Marine Transport & Management at the University of Newcastle in the UK. He was previously Professor and Head of the Department of Shipping & Transport Logistics at the Hong Kong Polytechnic University, Head of the Centre for International Shipping & Transport at Plymouth University, senior partner in his own transport consultancy company and Research Fellow at the University of Oxford Transport Studies Unit. He is a member of the Council of the International Association of Maritime Economists and sits on the editorial boards of *Transportation Research A, Maritime Economics & Logistics*, the *Annals of Maritime Studies* and the *Journal of Shipping & Logistics*. He is a Fellow of the Chartered Institute of Logistics & Transport and has been a transport adviser to the governments of Hong Kong, Egypt and the UK. He holds visiting professorships at a number of institutions and an Honorary Professorship at the University of Hong Kong.

Dr. Sharon Cullinane is an Honorary Fellow of the Centre of Urban Planning and Environmental Management in The University of Hong Kong where she recently spent three years as an Associate Professor. Having obtained a BA in Economics from Stirling University, UK in 1983 and a PhD in Transport Economics from the University of Plymouth, UK, in 1987, she has been working in the fields of transport economics and policy ever since. She has worked in various academic institutions around the world, including in the UK, Egypt and France.

Prof. Michael Firth is Chair Professor of Accounting and Finance at the Hong Kong Polytechnic University. He previously held academic appointments in Britain, New Zealand, and the US, and worked as an accountant and financial analyst in Britain. His research interests span the disciplines of accounting, finance, and industrial organization.

Dr Sophia Everett is Director of the Centre for Integrated Freight Systems Management, Melbourne Private University in Melbourne, Australia. She has held teaching and research appointments at Macquarie Graduate School of Management in Sydney, the Graduate School of Business and the Institute of Transport Studies at the University of Sydney. She was awarded a Doctoral Fellowship by the NSW Coal Association for research into transport infrastructure decision making and a post-doctoral fellowship into Australian flag shipping. In 1995 she was the recipient of the Australian Chamber of Shipping Research Award for investigation into the corporatization and privatization of Australian ports. She currently directs a continuing research programme into corporatization legislation and into rail access policy. Since 2004 she has managed the Permanent Secretariat of the International Association of Maritime Economists.

Stephen X.H. Gong is Lecturer in Finance at the School of Accounting and Finance of the Hong Kong Polytechnic University. His research interests include finance, maritime and transport economics, industrial organization and logistics. He has contributed to international refereed journals, books and book chapters, professional magazines and academic conferences in the areas of finance, shipping finance and logistics. He has also consulted with government agencies and large corporations in Hong Kong, China and the United States.

Prof. Hercules E. Haralambides is Professor of Maritime Economics and Logistics at Erasmus University Rotterdam and Director of the Erasmus Center for Maritime Economics and Logistics (MEL). He is also editor-in-chief of the quarterly *Maritime Economics & Logistics (MEL)*, formerly the *International Journal of Maritime Economics (IJME)*. Since 1989, he has been chairman of the Special Interest Group (SIG) on Maritime Transport and Ports of the World Conference on Transport Research Society. Among a number of consultative and advisory functions at national and international levels, he is an advisor to the European Commission, OECD, IMO, ILO and IBRD/IFC and one of the founding members of the International Association of Maritime Economists (IAME).

Dr Kwang Hee Kim graduated from the doctoral course of the Kobe University of Mercantile Marine.

Prof. Tae-Woo Lee is Visiting Professor at, among others, the Faculty of Economics and Politics in Cambridge, IMS at the University of Plymouth, the Hong Kong Polytechnic University, and Dalian Maritime University. Professor Lee was with Korea Maritime University from 1981 to 2003 and is currently vice-president of the Korea Marine Transport Policy Foundation (KMTPF).

Kunio Miyashita, PhD, born in 1943, has written *Market Behaviors in Competitive Shipping Market*, (1977) and *Global Competition of Physical Distribution Industry in Japan* (2002) He was Professor of International Physical Distribution at Kobe University from 1984 to 2004, Councillor to the Science Council of Japan 2003, and Professor of logistics at the University of Marketing and Distribution Science since 2004.

Prof. Enrico Musso is full professor in Transport Economics at the University of Genoa, where he coordinates the PhD in Transport Economics. The author of many books and essays in maritime economics, he is a member of the steering committees of *WCTRS-SIG2* (maritime transport and ports), of the international research network *TransportNET* and of the Italian Centre of Excellence in Logistics.

Dr Dong-Wook Song is Assistant Professor in Transport and Logistics Management at the University of Hong Kong. He is a Chartered Member of The Institute of Logistics and Transport, and holds visiting professorships at the Research Centre of Logistics, Nankai University, Dalian Maritime University, China and the Malaysia University of Science and Technology. He is interested in the managerial and strategic aspects of international logistics and maritime transport. Dr Song received a degree from the Korea Maritime University, and his MSc and PhD from the

University of Plymouth. He is an editorial board member of the *Journal of Shipping and Logistics* and the *International Journal of Logistics: Research and Applications*.

Dr Jose Tongzon is an Associate Professor at the Department of Economics, National University of Singapore (NUS). He specializes in trade and development focussing on the economies of Southeast Asia. He has taught trade and development-related courses at NUS and has written a number of articles in internationally refereed journals and books in the area of trade and development, including port and maritime issues with a special focus on Southeast Asia.

Prof. Alain Verbeke is associated with Templeton College (University of Oxford), the Haskayne School of Business (University of Calgary) and the Solvay Business School (University of Brussels, VUB). He has published 18 books and more than 160 other refereed publications, including several articles in the *Strategic Management Journal* and the *Journal of International Business Studies*. He is one of Europe's leading consultants on seaport policy and the strategic management of seaports.

Prof. Mariner Wang was Asian Region Sales Representative of Sea-Land Service Inc., Taipei, Taiwan in the period 1982–1986, then Sales Manager of C.Itoh Express, Taipei, Taiwan from 1987 to 1989. He received his PhD from Nagoya University (Japan), Graduate School of International Development in 1998 and from 2001 he has been the Professor and Associate Dean (International Affairs) of the Graduate School of Management (MBA), Ritsumeikan Asia Pacific University (Japan).

Dr Teng-Fei Wang is currently a post-doctoral research associate at the School of Marine Science and Technology, University of Newcastle upon Tyne. His research interests include performance measurement in the port industry, transportation network modelling and China's maritime and port economics and policies. Dr Wang received his BSc and MEng from Dalian Maritime University, PR China and his PhD in port economics from the Hong Kong Polytechnic University.

Ms Jian Hua Yang graduated from the doctoral course of the Kobe University of Mercantile Marine.

Prof. Jiaqi Yang is Professor and Deputy Dean at the School of Transportation, Wuhan University of Technology, PR China. He majored in Water Transportation Management at the Shanghai Maritime University. Professor Yang also holds a Masters Degree in Transport Management from the same university, as well as one from the World Maritime University where he majored in Shipping Management. Currently, he is finalizing his PhD at Erasmus University Rotterdam. He is the author of numerous papers in English and Chinese, as well as substantive reports on transport logistics for the Chinese Government.

Dr Shigeru Yoshida graduated from Kobe University of Mercantile Marine. His major areas of research include shipping and transport economics and management and he is now a professor of Kobe University.

Acknowledgements

The chapters in this book are further developed and more finely honed versions of papers that were originally delivered at the 2001 Annual Conference of the International Association of Maritime Economists (IAME) which was co-chaired by the editors and held in Hong Kong. We would like to express our deep appreciation of the organizational skill and effort of the chairman of the Local Organising Committee, Mr. C.K. Ng, and the conference secretary, Ms. Teresa Tong. In fact, in one way or another, all of the staff of the Department of Shipping & Transport Logistics at the Hong Kong Polytechnic University contributed immensely to what was widely considered to be a highly successful conference and a milestone in the development of IAME. Thanks are due to all of them for the tasks they undertook with their usual efficiency and enthusiasm. The various sponsors of the conference should also not be overlooked and, in that respect, the support of Hanjin Shipping and the IMC Group was particularly appreciated.

Prof. Kevin Cullinane owes a debt of gratitude to the School of Marine Science & Technology at the University of Newcastle upon Tyne for allowing him time and resources to undertake the work required to bring this book to fruition. He would also express his appreciation of the support afforded to him (largely during highly unsociable hours and moods) by Dr Sharon Cullinane, who also provided much valued professional advice on how to attack such a seemingly mammoth task. The helpful and efficient work of the editorial staff at Palgrave Macmillan, particularly Jacky Kippenberger, Katie Button and Nick Brock, has also been greatly appreciated. Finally, the editors would like to express their deepest appreciation of the efforts of the dedicated and talented individuals that have contributed chapters to this book. It is only fitting that it is their skill upon which the quality of this work relies.

Prof. Tae-Woo Lee
Prof. Kevin Cullinane

1
Introduction: Depth and Diversity in Analysing the Shipping and Ports Industries

Tae-Woo Lee and Kevin Cullinane

This book brings together an eclectic range of contributions from some of the world's leading maritime economists. Many have their origins in papers presented at the annual conference of the International Association of Maritime Economists (IAME) that was held in July 2001 in Hong Kong. Each of the contributions attempts to analyse at least a part of the present and future developments that are of interest to, or have an influence on, the world's shipping and ports industries. In so doing, a number of diverse generic issues and concerns are addressed, including competition, privatization, deregulation, alliances, efficiency, finance, pricing, logistics and information technology.

The contribution by Mariner Wang (chapter 2) is a discourse on the growth of container transportation in East Asia. It analyses the economic forces at play within this key region that constitutes the major driving force behind world shipping flows (especially in liner shipping but also, to a large extent, because of its important influence on bulk cargo flows). The author explains the emergence of the current situation in terms of different phases in the industrial and economic development of the East Asia region. Anyone with even the most limited appreciation of economic history in Asia will know that the initial phase in its modern economic development began with the rebuilding of Japan following the Second World War and the industrial revolution that occurred there in the late 1950s, 1960s and early 1970s. As Japan's industrial development approached maturity, the baton of fast-paced economic growth was taken up in turn by Korea in its efforts to recover from the ravages of a different war. With the vital support of American investment, Korea became the Asian economic powerhouse of the late 1970s and 1980s. As the pace of Korea's phenomenal growth began to flag, the Asian tiger economies (mainly Hong Kong, Taiwan and Singapore) stepped into the breach. In the mid-1990s, China emerged as the latest rapid developer in Asia, with most of the world's international trade today seeming to revolve around its influence. On the basis of the analysis contained within this chapter, the critical importance of the initial impetus provided by Japan cannot be underestimated. It has gone on to be one of the primary sources of foreign direct investment (FDI) in the nations that have come to subsequent industrial prowess. As time has gone by, of course, the influence of Japan over FDI in Asia has been supplemented by the later emergent economies, especially those of Korea and Taiwan. Against this context, this chapter analyses the nature of the container trades that move to, from and within the Asian region. It also highlights the role and position of the region's major container ports and the way in which both trades and ports are likely to prosper in the future. Particularly interesting in this respect is the part played in China by overseas investors, mainly from within Asia, and the benefits that may or may not accrue to these investments (see Panayides, Song and Nielsen, 2002).

Focusing on the liner shipping industry, Yoshida, Yang and Kim (chapter 3) point to the historical significance that cooperation between competitors has played within the sector. They provide details of the emergence, since 1995, of worldwide strategic alliances that have been forged between liner shipping companies and which have moved far beyond the boundaries of the more traditional cooperative arrangements that have historically existed between competitors in the liner shipping industry. The authors explain the motivation for entering into such alliances in terms of the network economies that they yield. Indeed, on the basis of the econometric analysis they undertake, the authors go so far as to suggest that the ultimate success of the partners within an alliance (in fact, therefore, the actual constituency of an alliance) is a function of the complementarity of existing networks and whether or not an alliance between the proposed members will yield the network economies that are sought. The econometric analysis conducted by the authors utilizes data from liner shipping services that are operated by the major Japanese liner shipping organizations (MOL, NYK and K-Line). The results graphically portray and explain the different changes in alliance membership over the last decade. Effectively, alliance membership is decided on the basis of mutual recognition of network complementarity and the scale of network economies that are derived from working together at the strategic level.

For the purposes of their analysis, Gong, Firth and Cullinane (chapter 4) define a 'financing method' as representing a joint decision on both governance structure and on available financing mechanisms. Based on this premise, the authors propose a new analytical framework for the analysis of financing methods that, it is suggested, should replace the more traditional and orthodox perspective that is usually adopted in neoclassical finance theory. A review of the alternative theories that seek to explain the corporate choice of capital structure leads the authors to conclude that an approach based on transactions cost economics (TCE) due to Williamson (1988) is likely to prove superior, in terms of explaining real-life behaviour in shipping finance decisions, to one based either on the capital structure irrelevance propositions (Modigliani and Miller, 1958) or on the pecking order theory (Myers, 1984). The authors go on to outline and advocate an approach to the analysis of financing methods in the shipping industry that fundamentally revolves around the TCE paradigm, but which is informed by a range of eclectic influences from other areas of economics and, most poignantly, by real-life behaviour. Specifically, they suggest that the TCE paradigm helps explain the combinations of capital structure and corporate governance decisions that are found to exist in the wider transport industry. The reason why this is the case is because asset specificity, the characteristics of the product (service) market under consideration (in particular, in the case of the shipping industry) and the nature of the seekers and providers of finance in this arena are all fundamentally transactional characteristics that exert a considerable influence on what decisions are taken.

Chapter 5, by Haralambides and Yang, addresses the controversial topic of 'flagging out'. The authors explain the motives behind the decisions of many of the world's shipowners to register their ships in a different country from where they are domiciled. Although a number of studies have revealed that there are a great many possible reasons which explain particular instances of this phenomenon, in most cases the major motivator is the desire to accrue cost savings, particularly in the area of crewing. The authors' perspective on flagging out is a novel one in that they focus specifically on the situation in mainland China. Perhaps surprisingly, this work reveals that flagging out is just as prevalent, and becoming increasingly so, in mainland China as it is amongst the more traditional maritime nations. The authors go on to apply fuzzy set theory to the derivation of an evaluation model of the flag choice decision in mainland China. In the absence of the imposition of a regulatory regime to prevent it from happening, the results suggest that the propensity for China's shipowners to flag out will inevitably increase as the country continues its inexorable and rapid economic development. In consequence, they advocate the imposition of suitable policies for reversing the trend towards flagging out. In particular, the authors propose that the Chinese government should take early steps towards the establishment of a 'second' or 'offshore' register as the major mechanism for deterring the overseas registration of mainland China's ships.

The contribution from Sophia Everett (chapter 6) examines the policy of deregulation that has been implemented within Australia and the impact that this has had upon the transport industry and the emerging intermodal environment. The aim of the policy was basically to enhance efficiency through promoting a more competitive environment. The nature of the policies introduced in order to stimulate competition and efficiency are outlined and their impact on Australia's intermodal transport is analysed. In particular, the author focuses on policy changes in the labour market, public sector reform and market deregulation. In covering the policy changes, mention is made of the peculiar governmental circumstances that exist in Australia; such that there are problems associated with policy implementation as the result of the sensitivities and conflicts that exist, and have to be overcome, between central and state governments, particularly one would imagine when the controlling powers are of different political hues.

Although not directly related to shipping and port issues, an important focus of Evesett's chapter lies with an analysis of the impact of deregulation in the Australian rail industry. This is, in fact, an aspect that is of crucial importance to the intermodal context in Australia because of the potential impact on the connectivity of the nation's ports.

The deregulation process and the competition instilled into the marketplace is presented as a prerequisite to, and facilitator of, the trend towards the increased intermodalism of Australia's freight transport industry.

One important conclusion drawn by the author is that deregulation has undermined and pulled down barriers to market entry and brought about competition in the rail market. This is graphically illustrated through the author's use of a case study of the 'Port Link' project.

A follow-up study on the medium-term impact of deregulation would be an interesting analysis to undertake in order to determine if there are any trends in market concentration, what long-term price elasticities have actually materialized (i.e. pricing tendencies) and what changes there have been in the quality of service provision. A further comparison of these findings internationally would be an interesting exercise; perhaps utilizing, for example, the rather well-established results of 'before-and-after' studies of the effects of transport deregulation in the United Kingdom in order to assess any degree of commonality in outcomes or to help explain differences.

Even where markets have seen no actual competition, attests Everett, they are at least now contestable post-deregulation and the benefits of this are being experienced in terms of greater efficiency, lower prices, better service quality etc. (see Baumol, Panzar and Willig, 1982).

In chapter 7 Mary Brooks seeks to find some initial answers to the question of whether good governance in ports is compatible with the desire of public sector interests to utilize ports as 'tools of economic development'. In other words, can port governance structures that have been designed in order to achieve commercial objectives meet the simultaneous expectations of government and the community, especially with respect to the economic development of the hinterland? In seeking a solution to this problem, the first thorny issue to grapple with is providing a definition of 'governance' that yields a common understanding. Once this has been achieved, the author moves on to focus specifically on the Canadian corporate governance code as providing a benchmark set of 'best practice' principles in governance and to analyse how best these can be implemented within any form of organization. In addressing the specific literature on port devolution, the author berates the current emphasis on full privatization and reminds us that this represents merely one end of a continuum of private sector participation in governance and only one of a whole range of possible characteristics that could be embodied in any given governance structure.

By contrasting the desired characteristics of a governance structure that facilitates economic development with those that are likely to be present when commercial considerations come to the fore, Brooks lays the groundwork for her ultimate message: We simply do not know enough about governance structures in ports and their relationship to the achievement of fundamental objectives of different stakeholders. There is a significant need for further empirical research into this very issue – an aspect that has motivated Mary Brooks into leading the formation of the Port Performance Research Network and the empirical work that it is about to undertake on a worldwide basis to answer just the sort of question that she poses with the title to this chapter.

Alfred Baird (chapter 8) provides an integrated analysis of the results of two surveys in order to illuminate the current state of the global ports industry with respect to the level and form of private sector involvement. The categorical result that, in the global arena, private sector participation in ports is fairly minimal may surprise some observers. However, while port privatization has formed a significant element in the transport and competition policies of certain nations (most notably the UK), there are many nations where the sector is regarded as so strategically, economically and militarily critical that private sector participation within it is political anathema. The author analyses which particular port activities have the most significant private sector involvement. Unsurprisingly, the private sector's most influential involvement is in the provision of stevedoring services within ports, with the government, or some other form of public institution, invariably financing and/or providing the land upon which ports are built. A public agency also invariably controls the waterways that yield access to the port. Baird goes on to investigate the motivation for privatization where it has occurred, the methods by which private sector ownership or control has been brought about and the advantages and disadvantages associated with greater private sector involvement. Clearly, the main motivation for privatization is a desire for greater efficiency in the provision of port services, but, as has been found to be the case in several other studies, there is an inevitable link to, and interaction with, labour reform within ports. In this, a proverbial 'Catch-22' situation prevails in many ports; while port labour reform may be one of the most desirable outcomes of a process of port privatization as a means of enhancing efficiency, where port labour reform has been implemented prior to any effort to privatize the sector, it is inevitably a much more attractive proposition for private sector investment than would otherwise be the case. One implicit conclusion that can be inferred from the content of this chapter is that, given the appropriate political will and context, there remains enormous scope within the worldwide port industry for further privatizations. These are likely to lead to enhanced levels of port efficiency across the globe and, as efficiency levels tend to equalize in the long term, to ever more intense competition between ports. This potential for further significant port privatization programmes is also likely to fuel the continuing vertical diversification of shipping lines into container handling via the investments they are continuing to make in dedicated and, sometimes, multi-user terminals. The continuation of the globalization policies of the major container terminal operators and the further concentration of the sector are also likely to figure prominently in the industry's future.

Following an elaboration of the varying definitions and terminology that are applied within research on hub ports and load centres, Song and Lee (chapter 9) provide a review of the relevant literature by classifying it into studies of port selection criteria, load centres and 'existing' hub ports. The authors argue that, although undoubtedly contributing to our understanding, previous research into the underlying determinants of how and why hub ports or load centres develop is not suited to explaining the contemporary shipping environment. The fundamental rationale underpinning their logic is the emergence of logistics and supply chain management as mainstream concepts in business and the consequent engulfing of the role of ports and transport generally within the wider framework of a supply chain or network. The authors quote the continuing emergence of third- and fourth-party logistics providers (3PLs and 4PLs) as categorical evidence for the radically different environment in which ports now compete to emerge as hub ports or load centres. This new environment manifests itself in many ways, but the proliferation of 3PLs and 4PLs mean, for instance, that the entities that make port choice decisions are changing again. These decisions had originally been the exclusive province of shippers, as the primary customers of ports. These rights have long since moved on to the liner shipping companies. Quite logically, the authors argue that a new era is emerging where these decisions are now increasingly being made by 3PL and 4PL integrators of the different entities within the supply chain and, although in many cases these companies have their roots in the provision of liner shipping services, this need not necessarily be the case. For this new era, it is argued, there is a new urgency in re-identifying the determining factors that bring about hub port or load centre status. Song and Lee then attempt to integrate their ideas with what they consider to be the results of previous research that remain germane to this new environment, to provide a new conceptual framework for underpinning any future empirical analysis of the determinants affecting port choice decisions.

In chapter 10, Jose Tongzon identifies the key factors that prompt the emergence of successful container transhipment hubs. It is argued that the factors that are most influential are: a strategic location (located on, or proximate to, the main East–West trade routes but, at the same time, inter-lining with either the major North–South trades or significant regional feeder networks), high levels of both port efficiency and connectivity, a solid infrastructure base supported by sophisticated IT systems and the offering of a wide range of port services that offer economies of scope and exhibit synergy. This is no meagre wish-list, with very few ports around the world having the benefit of all of these particular characteristics. Tongzon points to the examples of Rotterdam, Hong Kong and Kaohsiung, but selects Singapore as a case study to illustrate how and why it is this set of factors that have proven decisive in securing its position as the ultimate transhipment hub. Particular attention is paid to the role that has been played by the government of Singapore in supporting the achievements of its port. In so doing, the author highlights the effectiveness of the various government policies and incentives that have been put in place to support the activities and business of the port. Many of these have been prompted by the aspirations of Singapore to develop as an 'International Maritime Centre (IMC)', whose repute in maritime industries ventures far beyond its container port, to include prowess in bulk shipping (especially the tanker sector), bunkering, ship repair, shipping finance and insurance, ship regis-tration and, more recently, in the provision of third party logistics (3PL) services.

Finally, Tongzon concludes by analysing the increasingly competitive nature of the container port sector and the globalization strategy that Singapore's port operator has adopted in response to this challenge. The success of this strategy is guaranteed, it is argued, not only by its appropri-ateness, but also by the position of Singapore's container port as a leading incumbent in the market.

In chapter 11, Kunio Miyashita's chapter emphasizes the importance of ports developing their business and marketing strategies in recognition of their role as a node within international logistics networks. By focusing on the competitive position of the neighbouring ports of Kobe and Osaka in Western Japan, the author builds an econometric model of the trade volumes (distinguishing between imports and exports) that move through the two ports with destinations or origins within the Asian region. The key hypothesized determinants of these trade volumes are a mix of macroeco-nomic variables and factors specifically related to the services, as well as the market power, of each port. Elasticity measures for each of the determi-nants are obtained from estimating the model and implications are drawn concerning which macroeconomic factors have the greatest influence on the trade volumes of each of the two competing ports. Across a range of macroeconomic scenarios, the author analyses the competitive outcomes for the two ports under different port management strategies. He concludes that, in comparison to all other potential port management strategies, the optimum outcome would be achieved in a situation where the two ports were merged under a single management team and that the most appropriate logistics strategy would be one that was developed with the joint interests of the two ports in mind. However, Miyashita does acknowledge the political sensitivities involved in adopting such a strategy and provides us with an interesting insight into how political discussion on this proposal has progressed in Japan since it was first promulgated.

Should port financing and pricing constitute a valid policy option at the level of the EU? Within the boundaries set by such a policy, member states could have the flexibility to determine national ports policy and pricing specifics. Haralambides, Verbeke, Musso and Benacchio (chapter 12) review the possible basis upon which such a port pricing policy may be developed. From the perspective of the EU, it is critically important that port pricing policies are efficient and achieve a number of simultaneous objectives, the most notable of which in an EU context is to reflect (or recover) social opportunity cost and to stimulate a modal switch of freight movements from road to sea. From an EU perspective, the market cannot be the sole determinant of the strategic decisions (including those made on pricing) of ports. This is because of the relationship between port activities and wider social objectives.

The authors argue that while current EU policy advocates a strong commitment to a marginal cost pricing policy in support of achieving these objectives, little consideration is given to the practical implementation of such an approach (especially the collection of accurate and valid cost information). In contrast, the authors suggest that marginal cost pricing is neither a necessary, nor a sufficient, condition for achieving stated EU objectives for ports and that the collection of high-quality cost information would provide a much firmer basis for an appropriate EU-wide pricing policy.

Two distinct generic forms of port authority are defined in terms of the five parameters of ownership, objectives, autonomy, scope of activities and public resources allocated *à fonds perdus*. These parameters drive policy impacts and pricing policy.

A survey is conducted and the main findings show the importance of full cost recovery to ports and a majority support for the 'user pays' principle. The survey revealed criticism of current port pricing policies and advocated an even closer adherence to the cost recovery basis. Respondents baulked, however, at (inter)nationally, government-set standard prices.

Simulated testing of price elasticities revealed lower values for dry and wet bulk cargoes than for containers, general cargo and Ro-Ro. This also revealed a great diversity of values between ports in the Hamburg–Le Havre range (5 ports tested). These elasticity estimates have implications for the nature and form of pricing policy that may be implemented, insofar as they dictate outcomes to a large extent and these may cut across national member state interests.

Case studies of three ports in the UK and Ireland reveal subtleties in pricing despite the overall aim of securing full cost recovery. It was reported that changes in pricing policy would have a profound effect on the nature and value of their respective businesses and that it was perceived that, with the exception of landside congestion costs, most external costs had already been internalized by the investment made in complying with safety and environmental regulation.

Finally, Cullinane, Cullinane and Wang analyse the development of mainland China's container ports over the past decade – a time during which China has seen an explosion in its containerized trade. The authors adapt and apply a taxonomy, originally developed by Robinson (1998), that is based upon the position of a container port within a hierarchy of the type of liner shipping services that make most use of the port. This concept is utilized for explaining the phases in the development of the container port sector in mainland China and also for analysing the three geographical centres (in Southern, Central and Northern China) where container port competition is most intense. The influence of other factors is also considered. In this respect, the combined impact of China's port privatization policy, foreign direct investment in China's transport and logistics industry, the globalization policies of the world's major container terminal operators, planned transport infrastructure improvements and China's recent accession to the World Trade Organization (see Li, Cullinane and Cheng, 2003) are all critical to the future development of the sector.

References

Baumol, W., Panzar, J. and Willig, R. (1982) *Contestable Markets and the Theory of Industry Structure*, New York: Harcourt Brace Jovanovich.

Li, K.X, Cullinane, K.P.B. and Cheng, J. (2003) The Application of WTO Rules in China and the Implications for Foreign Direct Investment, *Journal of World Investment*, 4(2), 343–61.

Modigliani, F. and Miller, M.H. (1958) The Cost of Capital, Corporation Finance and the Theory of Investment, *American Economic Review*, 48, 261–97.

Myers, S.C. (1984) The Capital Structure Puzzle, *Journal of Finance*, 39, 575–92.

Panayides, Ph.M. Song, D.-W. and Nielsen, D. (2002) Foreign Direct Investment in China: The Case of Shipping and Logistics Corporations, *Proceedings of the International Association of Maritime Economists (IAME) Conference*, Panama, November.

Robinson, R. (1998) Asian Hub/feeder Nets: The Dynamics of Restructuring, *Maritime Policy and Management*, 25(1), 21–40.

Williamson, O.E. (1988) Corporate Finance and Corporate Governance, *Journal of Finance*, 43(1), 567–92.

2

The Rise of Container Transport in East Asia

Mariner Wang

2.1 Introduction

Since the Second World War, East Asia (consisting of Japan, the Asian newly industrializing economies (NIEs), the ASEAN 4 (Association of Southeast Asian Nations) and China) has enjoyed a remarkable record of high and sustained economic growth, growing faster than any other region in the world. East Asia's economic prosperity can be proven by its real GDP growth rate. In the period between 1980 and 1997, the Asian NIEs (Hong Kong, Singapore, Taiwan, Korea), the ASEAN 4 (Philippines, Malaysia, Thailand, Indonesia) and China have experienced average real GDP growth of 7.3 per cent, 5.9 per cent and 9.9 per cent respectively compared to an average 3.2 per cent worldwide, 2.7 per cent in Japan, and 2 per cent in the US and the EU (see Figure 2.1).

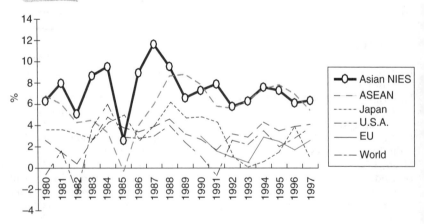

Figure 2.1   Real GDP growth rate of Asia in comparison to other countries/regions in the world
Note: Real GDP growth rates are calculated by simple average.
Source: IMF (1997).

The contrast is even more pronounced when the growth of per capita income across developing regions is compared. The high levels of economic growth in East Asia, particularly in the Asian NIEs, which has been showing a high real GDP growth rate in comparison with other regions or countries during the period between 1980 and 1997 is, in fact, closely related to the booming international trade being conducted regionally and globally, imparting to East Asia an extraordinary dynamism which greatly changed the shipping environment in the region.

East Asia's economic prosperity can be traced back to the September 1985 'Plaza Accord' by the Group of Five (the US, Germany, Britain, Italy, and Japan) intervening in the currency markets to drive the dollar significantly lower against the Japanese yen, expediting the second wave of Japanese enterprises' overseas forays in the Asia region. Other factors include the substantial appreciation of the Korean and Taiwanese currencies, labour shortages which have induced soaring wages since 1988, and also the drastic appreciation of the yen in the period 1991–1995 which also helped Taiwanese and Korean enterprises to survive in an era of increasing international competition. In response to the impending difficulties, Japanese, Taiwanese and Korean enterprises were forced to shift their production from the Asian NIES to ASEAN countries and China. As a consequence, the value of international trade and the ratio of trade reliance in East Asia have grown enormously, thereby generating a remarkable record of high and sustained economic growth unmatched by any other region in the world.

Triggered by this force, the volumes of shipping have risen steeply, generating a large concentration of container tonnage in East Asia. For years, ports in East Asia have been accounting for half of the world's top ten container ports (see Figure 2.2). Of particular note, taking year 1997 for instance, the container throughput of four hub ports in the Asian NIEs – Hong Kong, Singapore, Kaohsiung, and Pusan – accounted for almost a quarter of the global total.

The rapid economic growth experienced by East Asia was certainly not accomplished overnight. In the 1960s Japan became the focus of global attention as an emerging economic power that was rapidly catching up with the US and Europe. The economic growth in Japan was soon followed by the Asian NIEs (particularly, Taiwan and Korea), then ASEAN countries came to realize their huge potential for dynamic progress. Next, China started to take giant strides. In other words, in East Asia, one country after another has played the role of a forerunner, pulling the rapid economic growth of the entire region. This is the single most important factor behind the rapid upswing of East Asia on the global scene (Wang, 1999a).

This chapter falls into four main sections. The first explains the rise of international trade in East Asia. The second clarifies why East Asia has become the new emerging force in container transportation. The third illustrates developments in the main container ports in East Asia. Finally,

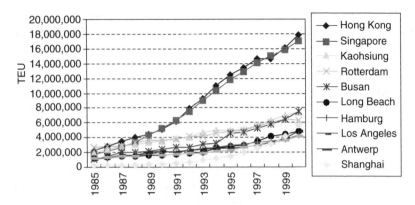

Figure 2.2   Annual throughput of the world's top ten container ports, 1985–1999
Notes: 1. The port of Shanghai has emerged as a new force since 1998.
2. Estimated figures in 2000 for the ports of Hong Kong, Rotterdam, Shanghai, Hamburg and Antwerp.
Sources: 1. *Containerization International Yearbooks*, 1988–2000, Emap Business Communications.
2. *Containerization International Yearbooks*, March 2001, Emap Business Communications.

the last section concludes by discussing the possible sustainability of container tonnage in East Asia into the twenty-first century.

2.2   The rise of international trade in East Asia

Trade is the oldest and most important economic nexus among nations. The modern interdependent world market economy makes international trade still more important.

International trade is an integral part of international business among individuals, firms, or other entities (both private and public), which consists of flows of both goods and services across national boundaries. Although flows of merchandise trade are tracked fairly accurately in most countries, such is often not the case for the international sale of services, where the diverse range of activities involved has limited the completeness and reliability of available data. Thus, this analysis focuses on the export and import of tangible goods – that is, merchandise trade flows.

International trade is usually the first phase in the international operations of a firm. Trade then leads to other modes of international operation: joint ventures, foreign direct investment (FDI) (see Rugman, Lecraw and Booth, 1985; Toyne and Walters, 1989), and licensing. International trade is a vital component of the economies of almost every country, the expansion of international trade is related to the economic growth of nations and the world economy itself, which leads to structural shifts in the economic organization of countries, providing new opportunities to firms, workers, and consumers. In short, the ability of a nation to seize export opportunities and respond to imports is a major determinant of its national economic performance.

The position of East Asia in world trade

As can be seen from Figure 2.3, the value of East Asia's exports as a percentage of world exports exhibited a steep upward curve from 14.2 per cent in 1980 to 26.2 per cent in 1995. However, the percentage suddenly dropped to 25.1 per cent in 1997 due to the financial crisis in Asia that year. In contrast, Figure 2.4 shows that during the same period, the value of East Asia's imports as a percentage of world imports increased from 14.7 per cent to 23.0 per cent.

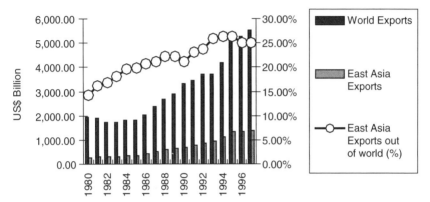

Figure 2.3 The value of East Asia's exports as a percentage of world exports, 1980–1997
Source: IMF (1998).

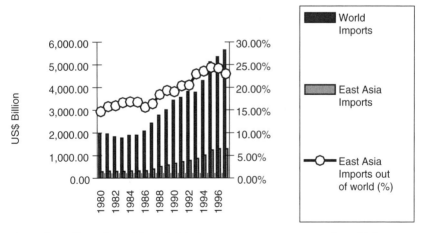

Figure 2.4 The value of East Asia's imports as a percentage of world imports, 1980–1997
Source: IMF (1998).

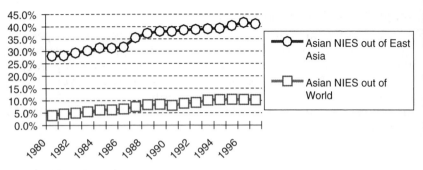

Figure 2.5 The value of Asian NIEs' exports as a percentage of East Asia/world exports, 1980–1997
Source: IMF (1998).

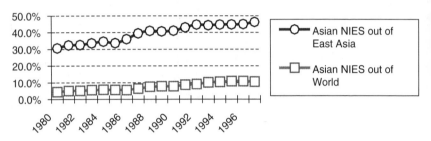

Figure 2.6 The value of Asian NIEs' imports as a percentage of East Asia/world imports, 1980–1997
Source: IMF (1998).

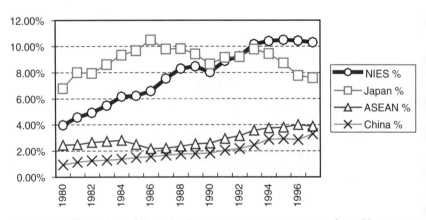

Figure 2.7 The value of East Asia's exports as a percentage of world exports, 1980–1997
Source: IMF (1998).

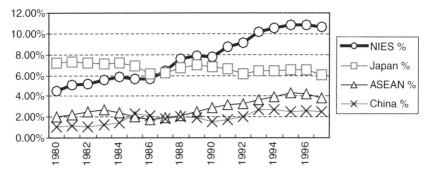

Figure 2.8   The value of East Asia's imports as a percentage of world imports, 1980–1997
Source: IMF (1998).

It is surprising to find that the value of the Asian NIEs' exports as a percentage of East Asia's exports climbed from 28.1 per cent in 1985 to 41.1 per cent in 1997. During the same period, the value of the Asian NIEs' imports as a percentage of East Asia's imports increased from 30.6 per cent to 46.2 per cent (see Figures 2.5 and 2.6).

Taking the value of Asian NIEs' exports/imports as a percentage of world exports/imports from 1980 to 1997 into consideration, it rose from 4.0 per cent to 10.3 per cent, and 4.5 per cent to 10.6 per cent respectively (see Figures 2.7 and 2.8).

It is also understandable that Japan has been taking the leadership in world trade. However, its weight in import value out of that of the world since 1987 (6.2 per cent) and export value since 1992 (9.2 per cent) has been overtaken by the Asian NIEs which stood at 6.5 per cent in 1987 on import value and 9.2 per cent in 1992 on export value (see Figures 2.7 and 2.8). The trend makes it clear that the scale of international trade in East Asia, particularly the Asian NIEs, has been on the rise.

Upswing in the scale of intra-regional trade in the Asian NIEs and the ASEAN four

Figures 2.9 and 2.10 illustrate the extent to which Asian NIEs' exports/imports relied upon which countries/regions during the period of 1980–1998.

It is evident from the above figures that since the mid-1980s, the percentage reliance of the Asian NIEs' exports/imports on the EU four (United Kingdom, Germany, France, Italy) and the US has been declining annually, while the percentage has been rising in respect of other East Asian countries, particularly within the Asian NIEs and ASEAN. This phenomenon can be attributed to three factors (Wang, 1999b); namely:

• Japanese enterprises' second wave of foreign direct investment in the Asian NIEs and ASEAN after the G5 agreement in 1985.

- Taiwanese and Korean enterprises' direct investment in ASEAN and China since 1988.
- The drastic appreciation of the Japanese yen against the US dollar during 1991–1995 (see Figure 2.11).

As has been already mentioned, the G5 agreement had forced the Japanese yen to appreciate drastically against the US dollar, accelerating the second wave of Japanese enterprises' foreign direct investment (FDI) and overseas production in the Asian NIEs (mainly Taiwan and Korea) and the ASEAN 4 since late 1985. Under this circumstance, Japanese enterprises have been positively exporting parts and plant equipment from Japan, by making use of raw material from the Asian NIEs and the ASEAN 4, to make semi-products or

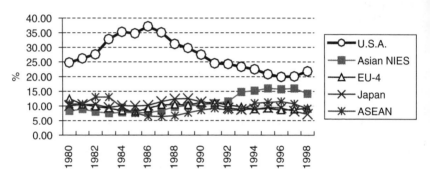

Figure 2.9   Percentage reliance of Asian NIEs' exports by countries/regions
Note: The EU four refers to United Kingdom, Germany, France, and Italy.
Source: Kaigai Keizai Deta, January 2000, Keizai Kikakuchou Investigation Bureau, Japan.

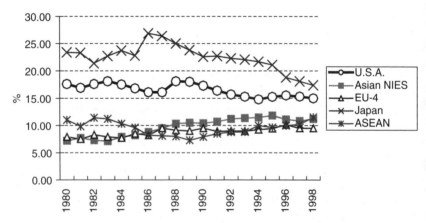

Figure 2.10   Percentage reliance of Asian NIEs' imports by countries/regions
Note: The EU-four refers to the United Kingdom, Germany, France, and Italy.
Source: Kaigai Keizai Deta, January 2000, Keizai Kikakuchou Investigation Bureau, Japan.

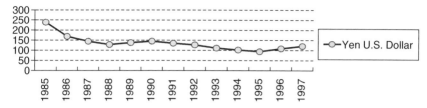

Figure 2.11   Foreign exchange rates of the Japanese yen against the US dollar, 1985–1997
Source: *Statistical Handbook of Japan*, 1998, Statistics Bureau, Management and Coordination, Agency Government of Japan.

complete manufactures for export; the former supplies the neighbouring affiliated plants located within the region, while the latter is exported directly to Japan, the US, Europe, etc. Furthermore, the governments of the Asian NIEs and the ASEAN 4 also successfully adopted policies of 'export-oriented industrialization' rather than policies of 'import substitution industrialization' which, combined with the Japanese enterprises' direct investments in East Asia, radically altered the structure of international trade inside and outside East Asia.

The increase in the percentage of import reliance on Japan for the Asian NIEs from 22.8 per cent in 1985 to 26.9 per cent in 1986 can be construed as the result of the export of plant equipment, parts etc., accompanied by Japanese enterprises' overseas production in the Asian NIEs (see Figure 2.10). On the other hand, in order to export completed manufactures back to Japan from the Asian NIEs, the percentage of the Asian NIEs' export reliance on Japan rose from 10.2 per cent in 1986 to 12.5 per cent in 1989 (see Figure 2.9). However, after 1988, the labour shortage in Korea and Taiwan triggered rises in labour costs, causing the export price to become uncompetitive internationally. To that end, many Japanese, Taiwanese and also Korean enterprises were forced to shift their production lines to the ASEAN 4 or China. As a consequence, the percentage of the Asian NIEs' export reliance on the ASEAN 4 rose from 6.7 per cent in 1988 to 7.8 per cent in 1989 (see Figure 2.9), demonstrating annual increases that reached 11.5 per cent in 1996. However, the figure dropped slightly to 10.7 per cent in 1997 due to the Asian crisis and, with the impact of this phenomenon still persisting, slipped to 9.1 per cent in 1998.

By contrast, due to the constant supply of raw materials and the procurement of semi-products from the ASEAN 4 to the Asian NIEs, the percentage of the Asian NIEs' import reliance on the ASEAN 4 has increased conspicuously since 1990, soaring to 11.5 per cent in 1998 (see Figure 2.10). Conversely, as a result, the percentage of the Asian NIEs' import/export reliance on Japan has been declining annually since 1987 and 1990, respectively hitting bottom at 17.4 per cent (see Figure 2.10) and 7.2 per cent (see Figure 2.9) in 1998.

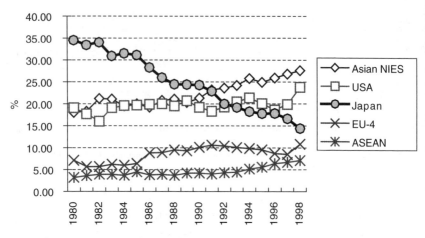

Figure 2.12   Percentage reliance of ASEAN's exports by countries/regions
Note: The EU-4 refers to the United Kingdom, Germany, France, and Italy.
Source: Kaigai Keizai Deta, January 2000, Keizai Kikakuchou Investigation Bureau, Japan.

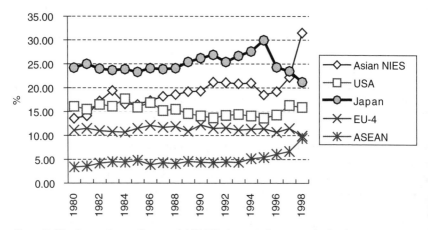

Figure 2.13   Percentage reliance of ASEAN's imports by countries/regions
Note: The EU-4 refers to the United Kingdom, Germany, France, and Italy.
Source: Kaigai Keizai Deta, January 2000, Keizai Kikakuchou Investigation Bureau, Japan.

In the 1980s, with the intention of procuring the supply of raw materials and semi-products manufactured locally for making completed manufactured goods for export, Japanese enterprises shifted their production bases to the ASEAN 4. As a result of this change of policy, the percentage reliance of the ASEAN 4's exports on Japan has been declining annually from 34.5 per cent in 1980 to 14.4 per cent in 1998 (see Figure 2.12). However, due to the supply of raw materials or semi-products to affiliated plants located in the region or the Asian NIEs, the percentage of the ASEAN 4's export reliance on the same region and the Asian NIEs has been increasing (with the exception of a drop with respect to the Asian NIEs in 1989). Direct investment from Japan, Taiwan and Korea into the ASEAN 4 also pushed their percentage of import reliance on Japan and the Asian NIEs to increase after the 1985 G5 agreement. For example, it was 21 per cent with respect to the Asian NIEs in 1994, and 30 per cent with respect to Japan in 1995 (see Figure 2.13).

Based on the above analysis, it is evident that because of the industrial policy shift of the Asian NIEs' from a policy of import substitution to one of export-oriented industrialization, the scale of international trade in the Asian NIEs has enlarged, expanding to East Asia as a whole. The sharp rise in international trade, in fact, is the basis of the surge in container tonnage in the region.

2.3 East Asia – the newly emerging force in container transportation

As was stated earlier, during the last decade, container tonnage in East Asia has been on the rise in parallel with the booming international trade expedited by the foreign direct investment from overseas, particularly from Japan, into the same region. This phenomenal change has been a matter of some concern throughout the world.

The surge of container tonnage in East Asia

Figure 2.14 illustrates the world's container traffic flow in 1990 and 1996. In 1990, the world's largest liner trade – the trans-Pacific (Far East/North America) service handled 5.34 million TEU. However, in 1996, traffic volume reached 8.21 million TEU, an increase of 54 per cent over that in 1990. On the other hand, during the same period the container volume handled in the Far East/Europe service, which is second to the trans-Pacific service in terms of container tonnage, was 2.89 million TEU. However, by 1996 traffic volumes reached 5.75 million TEU, a 100 per cent increase over that of 1990.

In 1990, container traffic volume handled in the intra-Asian (Japan, China, the Asian NIEs, ASEAN etc.) service was 3.5 million TEU, accounting for 15 per cent of the world's total container traffic volumes. In comparison, in the same year the container traffic volume handled in the trans-Pacific service was 5.34 million TEU and the volume in intra-European (EU) services was 4.55 million TEU accounting for 22.8 per cent and 19.4 per cent of the world's total container traffic respectively. It can be seen in Figure 2.14 that the scale of 1990s container traffic handled in the intra-Asian (Japan, China, the Asian NIEs, ASEAN) service could hardly compare with that in the trans-Pacific and the intra-European (EU) service. However, it surpassed both the traffic volume in

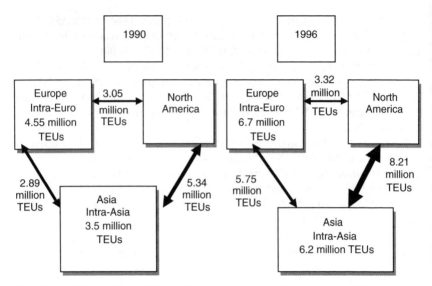

Figure 2.14   The position of intra-Asian services in world container traffic
Source: Nippon Yusen Kaisha Research Division.

the Far East–Europe services (2.89 million TEU) and the trans-Atlantic services (3.05 million TEU), where 12.3 per cent and 13 per cent of the world's container traffic were accounted for.

In 1996, the container traffic volumes in the intra-Asian service reached 6.2 million TEU, accounting for 16.6 per cent of the world's total container traffic. The figure could hardly rival the scale of the trans-Pacific services 8.21 million TEU that accounted for 22 per cent of the world's container traffic. However, it paralleled that of the intra-European service, and largely outpaced that of the Far East–Europe services (5.75 million TEU), as well as the trans-Atlantic services (3.32 million TEU), each of which respectively accounted for 15.4 per cent and 8.9 per cent shares of the overall world total. Clearly, the intra-Asian service, with its buoyant economic growth, has become the newly emerging force in world container traffic services (Wang, 1989).

Figure 2.15 indicates both the inbound and outbound container traffic volume handled by Asian countries (including Japan, the Asian NIEs, ASEAN 4 and Vietnam) in 1994. It is apparent from the figure that the shipping network in the intra-Asian region is quite complex. Additionally, the container traffic volumes differ by each individual trade lane. The main ports in Asia, (particularly, Yokohama and Kobe ports in Japan, Kwai Chung container terminal in Hong Kong, Tanjong Pagar, Keppel, Brani container terminal in Singapore, Kaohsiung port in Taiwan, Pusan port in Korea, Bangkok port and Laem Chang ports in Thailand) handled a

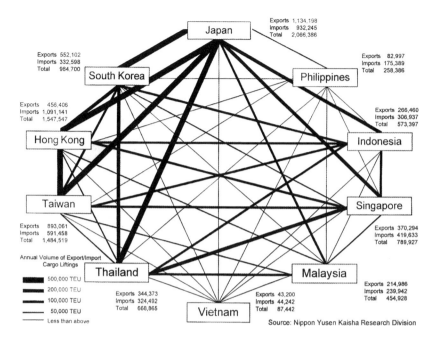

Figure 2.15   Intra-Asian container line service network and traffic volumes
Note: Figures were calculated from 1994 data.
Source: Containerization, No. 291, December 1996, Japan Container Association.

substantial volume of container tonnage that moved within the intra-Asian region.

It is surprising to find that the container traffic volumes handled by the ports in the Asian NIEs is very substantial if we focus on the container tonnage through the four hub ports in the intra-Asian services, instead of viewing it from its entirety. Figure 2.15 also shows that in 1994, the container traffic volume handled by the Asian NIEs reached 2.4 million TEU, which accounted for 54 per cent of container traffic in the intra-Asian service (excluding China).

Based on the above analysis, it can be construed that the surge of container traffic volumes in the intra-Asia trades contributed greatly to the hub ports in the Asian NIEs that dedicate themselves to facilitating the transhipment of container cargoes both from and to inside and outside the region.

The position of East Asia in global container tonnage

As has been already mentioned, container traffic tonnage in the intra-Asian service, particularly in the Asian NIEs, has been increasing substantially during the last decade. In this section, we consider the position of East Asia as well as the Asian NIEs in terms of container tonnage.

Table 2.1 The position of the ports of East Asia, the US and the EU 10 in total world container throughput (1,000 TEU)

Country	1985	1990	1991	1992	1993	1994	1995	1996	1997	1997/1985
Japan	5,517	7,956	8,782	8,965	9,349	10,417	10,604	11,033	10,892	1.97
(%)	9.9	9.3	9.4	8.7	8.3	8.1	7.7	7.3	6.7	
Port of Hong Kong	2,289	5,101	6,162	7,972	9,204	11,050	12,550	13,460	14,567	
Port of Singapore	1,699	5,224	6,354	7,560	9,046	10,399	11,846	12,944	14,135	
Port of Kaohsiung	1,901	3,495	3,913	3,961	4,636	4,900	5,053	5,063	5,693	
Port of Pusan	1,148	2,348	2,571	2,751	3,071	3,826	4,503	4,725	5,234	
Asian NIEs Hub Ports	7,037	16,167	19,000	22,244	25,957	30,175	33,952	36,192	39,629	5.63
(%)	12.6	18.9	20.3	21.6	22.9	23.5	24.7	24.0	24.2	
Taiwan	3,075	5,451	6,130	6,179	6,795	7,310	7,849	7,866	8,516	
Korea	1,246	2,348	2,571	2,751	3,071	3,826	4,503	5,078	5,637	
Asian NIEs	8,309	18,124	21,271	24,462	28,116	32,585	36,748	38,995	42,452	5.11
(%)	14.9	21.2	22.7	23.8	24.8	25.4	26.8	25.9	25.9	
Philippines	638	1,408	1,441	1,158	1,663	2,016	1,892	2,336	2,507	
Thailand	400	1,078	1,172	1,337	1,492	1,772	1,962	2,052	2,100	
Indonesia	229	924	1,153	1,397	1,611	1,912	2,048	1,764	1,920	
Malaysia	389	888	1,074	1,218	1,398	1,746	2,075	2,550	2,976	
ASEAN 4	1,656	4,298	4,840	5,110	6,164	7,446	7,977	8,702	9,503	5.74
(%)	3.0	5.0	5.2	5.0	5.4	5.8	5.8	5.8	5.8	
China	446	1,204	1,506	2,011	2,785	4,064	4,682	5,238	5,788	12.98
(%)	0.8	1.4	1.6	2.0	2.5	3.2	3.4	3.5	3.5	
East Asia	15,928	31,582	36,345	40,548	46,414	54,512	60,011	64,321	69,038	4.33
(%)	28.5	36.9	38.8	39.3	41.0	42.5	43.7	42.7	42.2	

Table 2.1 The position of the ports of East Asia, the US and the EU 10 in total world container throughput (1,000 TEU) – *continued*

Country	1985	1990	1991	1992	1993	1994	1995	1996	1997	1997/1985
U.S.A.	11,533	15,245	15,546	16,889	17,390	18,442	19,104	21,777	23,758	2.06
(%)	20.6	17.8	16.6	16.4	15.4	14.4	13.9	14.4	14.5	
EU10	14,782	19,697	20,848	22,134	22,872	25,046	26,846	28,848	33,187	2.61
(%)	26.4	23.0	22.3	21.5	20.2	19.5	19.6	19.1	20.3	
Others	13,660	19,073	20,907	23,335	26,536	30,320	31,278	35,807	37,761	3.47
(%)	24.4	22.3	22.3	22.7	23.4	23.6	22.8	23.8	23.1	
World	55,903	85,597	93,646	102,906	113,212	128,320	137,239	150,753	163,744	2.93

Note: The EU10 refers to the United Kingdom, Germany, France, Holland, Italy, Spain, Belgium, Portugal, Greece, and Denmark.
Source: Containerization International Yearbook, 1983–1999, Emap Business Communications.

Table 2.1 demonstrates the container throughput by port, region and country as well as the shares in comparison to the world's total container traffic from 1985 to 1997. As is evident from the table, during this period, container throughput in East Asia rose from 15.9 million teu to 69.0 million TEU, respectively accounting for 28.5 per cent and 42.2 per cent of the world's total container tonnage.

Looking more minutely into the hub ports of the Asian NIEs, it can be found that during the same period the share of ports in the Asian NIEs out of the world's total container throughput rose from 14.9 per cent to 25.9 per cent. It is also surprising to find that, during the same period, the percentage of total container throughput at the Asian NIEs' main ports – i.e. Hong Kong, Singapore, Kaohsiung and Pusan – rose from 12.6 per cent to an astounding 24.2 per cent, accounting for approximately one quarter of the world's total container tonnage. In comparison, ports in the ASEAN 4 rose from 3.0 per cent to 5.8 per cent; ports in China rose from 0.3 per cent to 3.5 per cent; while the US declined from 20.6 per cent to 14.5 per cent and the EU 10 declined from 26.4 per cent to 20.3 per cent. It is apparent that these hub ports in the Asian NIEs have made a major contribution to the substantial growth of container traffic tonnage in East Asia as a whole (Wang, 1997; Wang, 2000).

Figures 2.16 and 2.17 indicate the volume and percentage of container traffic handled by East Asia (Japan, the Asian NIEs, ASEAN 4, China), the US and the EU 10, as well as their position in the world's total container traffic volume during the period 1980–1998.

As can be seen from Figure 2.16, during the period presented, the volume of container traffic handled by East Asia, particularly the Asian NIEs, has been demonstrating conspicuous annual increases. In 1980, the EU 10, East

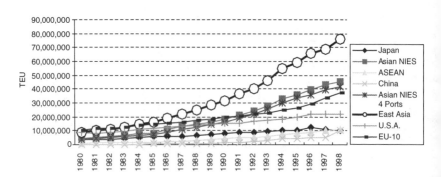

Figure 2.16   The position of East Asia, the US and the EU 10 in world container traffic volumes, 1980–1998
Note: The EU10 refers to the United Kingdom, Germany, France, Holland, Italy, Spain, Belgium, Portugal, Greece, and Denmark.
Source: *Containerization International Yearbook*, 1983–2000, Emap Business Communications.

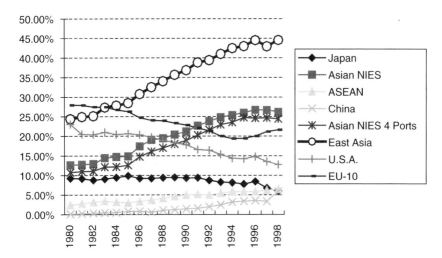

Figure 2.17   Volume shares of East Asia, the US and the EU 10 in world container traffic, 1980–1998
Note: The EU 10 refers to the United Kingdom, Germany, France, Holland, Italy, Spain, Belgium, Portugal, Greece, and Denmark.
Source: *Containerization International Yearbook*, 1983–1999, Emap Business Communications.

Asia, and the US were the top three container traffic regions in the world. However, in 1984 the container traffic volume of East Asia (14.84 million TEU) began to surpass that of the EU 10 (14.31 million TEU), and has been increasing drastically ever since to become the hub of the world's container concentration, unmatched by any country or region in the world.

In contrast, during the same period, the traffic volume in the EU 10 and the US (10.9 million TEU) has been demonstrating slow growth, ranking second and third in container tonnage respectively. As a consequence, the gap in container traffic volume between East Asia and the EU 10, as well as the US, has become ever larger since 1984. In 1995, the differences in container traffic volumes between East Asia and the EU 10, as well as East Asia and the US, were among the largest, with 32.22 million TEU and 39.52 million TEU respectively. This was unprecedented in maritime history. As a result, the percentage share of East Asia in the world's total container traffic volume reached 43.8 per cent in 1995, compared to a comparable figure of 24.4 per cent in 1980.

Furthermore, special attention should be paid to the shares of the Asian NIEs. In 1986, they demonstrated a drastic increase of 27.7 per cent over the previous year, in which the G5 agreement expedited Japanese enterprises' overseas production (labour-intensive industry) in the Asian NIEs. Thus, by exporting plant equipment, parts etc., substantial container concentration was generated in the region. While in comparison, during the same period, the share of the EU 10 and the US dropped from 30 per cent to 19.9 per cent, and from 23 per cent to 14.5 per cent respectively.

From this phenomenal change, it can be construed that the centre of global container traffic has been shifting from Europe and the US to East Asia (See Table 2.1).

2.4 Developments in the main container ports in East Asia

Through the early 1970s, Japan consolidated its economic and trade position, spurring shipping services to expand substantially. By contrast, it has been the emergence of East Asia, particularly the Asian NIEs from the 1960s through the 1970s and 1980s to the present, and more recent rapid economic growth in Southeast Asia, that underlie containerization and the development of container handling capacity in the region's ports. Over the last decade, the shipping market in East Asia has emerged as an exceptional generator of container traffic. The Asian NIEs are among the most successful industrialized countries in the developing world. The rapid and sustained growth of international trade and container tonnage in East Asia, particularly Hong Kong, Singapore, Kaohsiung and Pusan as the hub ports, can be attributed to the foreign direct investment boom, particularly from Japan, into this region.

The port of Hong Kong, as a major mainline or regional transport mega-hub, dominates the central cluster, and its continuing high growth reflects the rapid economic development in southern China, as well as its central position and significance. Given its business-friendly environment and world-class infrastructure, the port of Hong Kong is not only blessed with a unique geographical location bearing mainland China as the hinterland, but also plays a vital role as the entrepôt for container transhipment for both Asia/North America and Asia /Europe traffic lines. Additionally, it provides the feeder services for the export and import cargoes between inland China and the adjacent Pearl River Delta. For these reasons, the port of Hong Kong has been the leading container port in the world for many years.

Growth at the port will be boosted by the time the US$2 billion container terminal nine (CT9) comes on stream. The first of the six berths came on-stream in May 2000. Once complete, the new facility will occupy an area of 70 hectares of reclaimed land and will consist of four deep-sea berths and two feeder berths. The new terminal, to be operated by existing Kwai Chung operators, Hutchison International Terminals, was operational towards the end of 2001 having a design capacity of 2.6 million TEU, although the actual throughput is expected to be much higher when efficiency measures kick in.

The port of Singapore is the load centre of container ports in Southeast Asia. The focus of expertise in the port of Singapore has mainly been in the

fields of containerized freight and logistics. As a maritime hub encompass-
ing the whole range of services, Singapore, with ongoing liberalization of
financial and telecommunication services, will be an even bigger attraction
for more carriers to call (*Asiaweek*, 1997; *New Straits Times*, 1998). The port
of Singapore is currently under the management of PSA (Port of Singapore
Authority) Corporation. Initiated in 1996, PSA's globalization efforts have
now become a significant part of the group's business. Since embarking on
its first project in Dalian, China in 1996, up to 1999, PSA has been involved
in developing, managing and operating ten ports in seven countries
(Negara Brunei Darussalam, China, India, Italy, Portugal, South Korea,
Yemen). Total throughput from these overseas projects and joint ventures
stood at 1.72 million TEU in 1999, bringing in combined revenue of over
400 million Singapore dollars. In 1999 alone, PSA was invited to develop
container terminals in Muara, Negara Brunei Darussalam; Sines, Portugal;
and Inchon, South Korea. Additionally, PSA's Greenfield project at Aden,
Yemen has also been launched into full-scale operation. These projects are
expected to contribute significantly to PSA's growth and have attracted
international attention to the group (see Table 2.2).

Moving ahead, PSA will vigorously pursue overseas expansion and
the development of its IT (Information Technology) services. In 1999, PAS
and Hong Kong's China Merchants Holdings (CMH) Group formed a new
logistics company in China, the China Merchants-PSA Logistics Network
Co. Ltd. Aiming to be a fully fledged logistics provider, the company has
since secured a number of logistics contracts with multinational companies
and their representatives (such as Philips Lighting and Tiat Trading, the
distributor for Heineken Beer and Evian Natural Spring Water) in major
cities including Shanghai, Guangzhou and Tianjin. It is actively looking
for prospective partners to set up regional logistics centres so as to build up
a pan-China logistics network.

On the technology front, the year 2000 saw PSA Corp spin off its IT/
e-commerce arm into a separate subsidiary company, Portnet. Com Pte Ltd.
is based on PSA's Portnet system, a nationwide e-commerce network that
involves the entire shipping and port community operating through
Singapore.

PSA has also teamed up with P&O Ports to acquire a stake in P-Serve
Technologies, a provider of Internet-based track and trace solutions. Under
an agreement signed by the three parties, PST will provide the infrastructure and project management while PSA and P&O will market the solution,
e-Logicity, to their customers through their respective ports and logistics
business (*Lloyd's Maritime Asia*, 2000a).

The port of Kaohsiung, with its geographical advantage located along
the southwestern coast of Taiwan on the key trade lanes running through
the Taiwan Strait and the Bashi Channel, is the largest international
seaport in Taiwan and also an ideal hub port for transhipment of the

Table 2.2   PSA's overseas container terminal investments

Container terminal	Aden	Muara	Sines	Inchon
Country	Yemen	Negara Brunei Darussalam	Portugal	Korea
No.of Berths	2	1	3	3
Total Quay Length	Phase 1: 1,700 m	250 m	940 m	900 m
Draft	16.0 m	12.5 m	16.0–17.0 m	12.0–13.0 m
Stacking Area (m^2)	350,000	62,000	—	—
Container Handling Equipment	4 post panamax quay cranes 8 yard cranes	2 quay cranes 5 reach stackers	—	—
Terminal Capacity (TEUs)	500,000	220,000	1,4 00,000	—

Container terminal	Dalian	Fuzhou Quingzhou	Fuzhou Aofeng	Voltri
Country	China	China	China	Italy
No. of Berths	4	2	1	4/2*
Total Quay Length (m)	1,113	519	156	1,400
Draft (m)	12.0–14.0	11.7	6.5	15.0
Stacking Area (m^2)	392,000	66,000	6,300	350,000
Container Handling Equipment	7 quay cranes 23 yard cranes	quay cranes 8 yard cranes	1 fixed crane 2 reach stackers	8 quay cranes 13 yard cranes
Terminal Capacity	1,500,000 teus	340,000 teus		870,000 teus

Container terminal	Venice	Tuticorin	Pipavav
Country	Italy	South India	India
No. of Berths	3	2	2** /1***/1****
Total Quay Length (m)	510	370	975
Draft (m)	9.0	11.9	13.5
Stacking Area (m^2)		55,000	100,000
Container Handling Equipment	4 quay cranes 8 yard cranes	2 quay cranes 4 rubber-tyred gantry cranes	2 quay cranes 5 yard cranes for phase 1
Terminal (TEUs) capacity	300,000	200,000	200,000

Source: PSA Corporation.
Note: *Denotes car carrier and Ro-Ro Berth
          **Denotes general cargo berth
          ***Denotes bulk berth (to be converted to container berth in phase 1)
          ****Denotes liquid berth.

export cargoes between the west coast of North America and southeast countries. Furthermore, owing to the opening of direct sailing between the port of Kaohsiung and the ports of Xiamen and Fuzhou (Fujian Province) in China in 1997, the port of Kaohsiung has been provided with more room for securing transhipment cargoes from North America and China. In the past two decades, the port of Kaohsiung has made significant contributions to Taiwan's economy. For instance, the Export Processing Zone (EPZ), situated on the Chungtao Peninsula in the harbour, is a huge complex of factories and offices that brings in raw materials and parts duty free for manufacturing and export. In addition, the port of Kaohsiung provides one of the world's largest ship scrapping facilities. Of particular note is the privatization of stevedoring services in 1998, which is regarded as one of the reasons for the double-digit growth (11.4 per cent) in container throughput, reaching 6,985,361 TEU in 1999 and 7,425,831 TEU in 2000 (Chen, 1998).

Table 2.3 provides data illustrating the double-digit growth in inbound/outbound vessel numbers, container throughput at the container terminal and cargo throughput (container + general cargo), as well as the hourly handling capability at the conventional terminal after the privatization of stevedoring services in the port of Kaohsiung.

Table 2.3 The comparison in vessel inbound/outbound, cargoes throughput before/after the privatization of stevedoring services at the port of Kaohsiung

Items		*Unit*	*Performance during January–March 1998*	*Percentage increase over January–March 1997*
Vessels	Total	Number	8,312	+13.09
		Tonnage	140,987,381	+11.33
	Inbound	Number	4,160	+13.29
		Tonnage	70,662,313	+11.17
	Outbound	Number	4,152	+12.89
		Tonnage	70,325,068	+11.49
Container Cargo	Throughput	TEU	1,401,739	+12.85
Container + General Cargo	Throughput	Ton	28,447,909	+20.36
Conventional Terminal	Throughput	Ton/hour	98.5	+13.00

Note: MPH = Movement Per Hour.
Source: Kaohsiung Port Bureau.

The port of Pusan, lying on the southeast coast of the Korean peninsula facing the Korea Strait, plays a pivotal role as the major port of the regional centre in Northeast Asia, bridging the export and import cargoes between the ports of Vladivostok and Vostochy in Russia, the ports of Dalian, Qingdao, Lianyungan etc. in north China, as well as ports facing the East Sea (Sakata, Akita, Niigata, Toyama etc.), North America, Europe as well as Southeast Asia. This geographical advantage has allowed the port of Pusan to join the league of the world's top ten container ports.

The 2000 increase of 17.1 per cent in containers handled in South Korea jumped by its highest margin since the 1995 Kobe earthquake, when South Korea won a big piece of Japan's transhipment cargo. Container through-put is expected to increase annually because South Korea's role as a logistics hub for north Asia is becoming more and more viable.

Additionally, the Korean government, with a view to attaining balanced national land development and to coping smoothly with the drastic increase of container volumes resulting from brisk Korean economic growth, consigned the Korea Container Terminal Authority (KCTA) to launch intensive short-, medium- and long-term port development projects in 1994. Those projects incorporated into the medium-term timescale include port development and expansion projects in Gamman port (known as Pusan's fourth phase port development (1991–1997), an unprecedented case in the history of port development in Korea as the Korean government invited capital injection from the private sector – the carriers); the port of Kaddo–Pusan New Port (2001–2007); the port of Kwangyang (first phase: 1987–1997, second phase: 1995–2003), located about 50 kilometres and 170 kilometres to the west of Pusan city respectively and the Southern port of Inchoen (2001–2003). Long-term projects include the third phase (1999–2008) and the fourth phase (2002–2011) of the port of Kwangyang. Upon completion of the second phase development (4 berths for 50,000 dwt vessels and four berths for 20,000 dwt class vessels) in 2002, the port of Kwangyang has the capacity to handle 2.4 million TEU annually (Wang, 1998).

The first two phases of the construction of Pusan New Port require 3.70 trillion won and 2.45 trillion won respectively. Exclusively for con-tainers, the consortium of builders and operators includes Samsung (27.5 per cent), Hyundai (16.5 per cent), Hanjin (12.5 per cent) and KCTA (9.0 per cent). The consortium will construct a container terminal capable of accommodating seven 50,000 dwt and three 20,000 dwt container vessels in the first phase development, as well as eight 50,000 dwt and six 20,000 dwt container vessels in the second phase development, together with back-up facilities. In comparison, the construction of third-phase and fourth-phase development requires 844.6 billion won, and has the capability of accommodating ten 50,000 dwt container vessels. In the year 2011 when the long-term Pusan port development plan is completed, there will be 65 berths, with annual container handling capability of 15.75 million TEU. Another upgrade for the entire Pusan port is to spend some 58 billion won deepening the quayside channel to 15 metres from the current 13 metres.

As South Korea adds container capacity, this will fuel the preference for container lines to use Pusan as a hub in order to avoid expensive Japanese facilities and also access the increasingly attractive mainland China market.

In addition, the restoration of rail links between North and South Korea triggered by the Seoul–Pyongyang rapprochement in 2000 will be an opportunity for both South and North to make another great leap forward. This, coupled with the above mentioned ambitious port expansion plans, will strengthen the Korean Peninsula's claim to be a major hub of Northeast Asia (*Lloyd's Maritime Asia*, 2000b).

Special attention should be paid to the ports in China. In recent years, Chinese exports have moved into high gear to supply the needs of swelling world trade, and the port of Shanghai is a large beneficiary from the huge expansion in trade to the East and West Coast of the US.

Container throughput at Shanghai was surprisingly resilient in 1998, handling 3.05 million TEU in 1998, up 21 per cent on the previous year's 2.53 million TEU, placing Shanghai tenth in the world container port rankings; the first time in recent memory that the port was placed in the global top ten largest ports (see Figure 2.18).

Despite growing concern that operational capacity is being squeezed, Shanghai continues to occupy a lofty perch, way ahead of the Chinese pack. In 1999 and 2000, it demonstrated a 37.2 per cent and 33.4 per cent increase in box traffic to 4.2 million TEU and 5.6 million TEU, placing Shanghai port seventh and sixth in the world rankings respectively. The remarkable nature of this achievement is the result of the sustained investment in new terminals

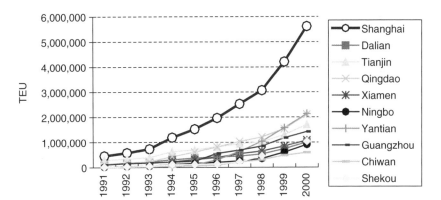

Figure 2.18   Container throughput at the major Chinese ports
Source: Ministry of Transportation, China.

and the introduction of world-class port management expertise in Hutchison Port Holdings to manage Shanghai Container Terminals (SCT). These factors have driven the rapid ascent of Shanghai port through the world rankings.

Furthermore, throughput at the South China port complex of the Shenzhen district – including the ports of Shekou, Yantian and Chiwan – also cannot be ignored. In 1999, they posted a huge 53 per cent rise in throughput to 2.98 million TEU. In 2000, containers transported in and out of the Shenzhen district reached 3.99 million TEU – up 33.8 per cent on the previous year. This pushed the Shenzhen complex comprising Shekou Container Terminals (SCT), Kaifeng Terminal (KFT) and Yantian International Container Terminals (YICT) into the top 11 of the world's container ports in 2000. The Shenzhen ports are expected to handle around 5.2 million TEU in 2001. Undoubtedly, Shenzhen will continue to be a major challenger for Hong Kong's market share.

Additionally, being attracted by the buoyant economic growth in southern China as well as the opening of direct sailing between Taiwan and China as aforementioned, in 2000 Maersk-Sealand revamped its China–Europe routes to include the port of Xiamen and Fuzhou, as it expanded capacity in line with the strong growth in trade. The ports of Xiamen and Fuzhou, which are located across the Straits of Taiwan, benefited from a mix of new direct call liner services to/from Europe and North America and the growing relay trade to/from Taiwan. The container throughput of these two ports increased by 30.7 per cent and 28.0 per cent respectively in 1999 and 27 per cent and 25 per cent respectively in 2000 (*Containerization International*, 2000). On the basis of China's continuing economic development and its admission into the World Trade Organization (WTO), the future is expected to be just as bright for the nation's ports.

2.5   Conclusion

The concentration of container tonnage in East Asia and the intensity of operations are significant, not only in regional but also in global terms. Nor is it the simple magnitudes involved that make regional concentration significant. It is also the way in which the ports are linked together into global and regional shipping networks. The mainline hub/feeder structure focuses on large flows of containers and shipping capacity on to a small number of extremely efficient ports; this, combined with the enhanced throughput capacity of these ports, gives East Asia particular significance at both a global and a regional scale.

1999 saw the economic rebound in Asia long after the financial crisis originating from the Thai baht devaluation in July 1997. The period of economic crisis-induced stagnant growth has come to an end. The Asian region has generally turned the corner and is now heading towards economic recovery, though the overall rate of growth is set to slow. World economic growth and world trade are forecast to improve further in 2000, with estimates for world trade growth at 7.1 per cent against 4.5 per cent in 1999. Given that a sustained economic cyclical upswing is now underway, this robust performance can be attributed to the acceleration of the Asian economic recovery plus an upturn in trans-Pacific shipments in the second half of 1999.

According to information available from Containerization International Market Analysis, global container output, after remaining largely static for several years, increased by 25 per cent in 2000 to a new record high of more than 1.85 million TEU (see Figure 2.19). The rise in container production was firmly underpinned by improved global trade growth during 2000 and a record delivery of over 600,000 TEU of additional vessel slots in the same year. Additionally, the world fleet of containers was forecast to grow by 40 per cent over the next four years, to reach 20.9 million TEU by 2005, according to *Containerization International* (2001). This represents a year-on-year growth of 8.5 per cent, and is slightly slower than the 11 per cent rise in the fleet that occurred in 2000.

Figure 2.20 illustrates the forecast real GDP growth in East Asia from 2000 to 2002, surveyed by DB Global Markets Research. It reveals that the East Asian real GDP growth in 2000 outside Japan increased by 4.0 per cent in the Philippines, 8.0 per cent in China, 9.0 per cent in South Korea, 9.5 per cent in Singapore and 9.8 per cent in Hong Kong, signalling an upward revision of economic expansion in the region. However, East Asian real GDP outside Japan and China will increase by about 5 per cent in 2001, down from 7 per cent last year, and China will likely turn in about 7 per cent. The reason for the regional slide is due to the factor that there just isn't enough domestic demand to offset the feared export slump, as consumer spending in Asia is weakening.

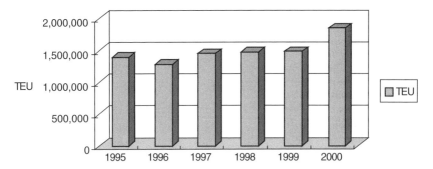

Figure 2.19　World container output, 1995–2000
Note: Totals include maritime and regional container types.
Source: *Containerization International*, February 2001, Emap Business Communications.

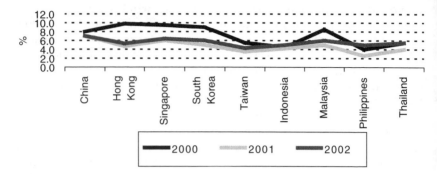

Figure 2.20   Forecast real GDP growth in East Asia, 2000–2002
Source: DB Global Markets Research.

However, Asia is more resilient and its vulnerability to global decelera-
tion is fairly limited, though Japanese GDP growth is forecast to slow from
2.0 per cent in 2000 to 1.5–1.8 per cent or even to as low as 1.0 per cent in
the year to March 2002.

As predicted by Ocean Shipping Consultants, the share of East Asia in the
world container market is set to rise from 43.7 per cent in 1995 to 50.2 per
cent by the year 2000 – despite strong growth in other world markets, and
by 2010 its share is forecast to rise to around 52.4 per cent – equating to a
massive 244 million TEU, making it the world centre of highest intensity
for liner shipping operations. In that decade, China and the second tier
of industrializing countries will really begin to make their mark (*Lloyds
Maritime Asia*, 1996).

References

Asiaweek (1997) Malaysia vs. Singapore: A New Spin in the Anti-complacency Drive,
    17 January, 20.
Chen, S.-L. (1998) The Privatisation of Port of Kaohsiung, *Proceedings of 1998
    International Conference on Shipping Development and Port Management*, KaoPort 21,
    Taiwan, 26–29 March.
Containerization International (2000), March, Emap Business Communications.
Containerization International (2001), March, Emap Business Communications.
IMF (1997) *International Financial Statistics Yearbook*, Washington, DC: The International
    Monetary Fund.
IMF (1998) *International Financial Statistics Yearbook*, Washington, DC: The International
    Monetary Fund.
Lloyd's Maritime Asia (1996), March, LLP Limited, 3.
Lloyd's Maritime Asia (2000a), December, LLP Limited, 24–6.
Lloyd's Maritime Asia (2000b), October, LLP Limited, 6–7.
New Straits Times (1998) Efforts to Increase Use of Port Klang, 17 July, 8.
Rugman, A.M., Lecraw, D.J. and Booth, L.D. (1985) *International Business: Firm and
    Environment*, New York: McGraw-Hill.

Statistical Handbook of Japan (1998) Statistics Bureau, Management and Coordination, Agency Government of Japan, Tokyo.

Toyne, B. and Walters, P.G.P. (1989) *Global Marketing Management: A Strategic Perspective*, Boston: Allyn and Bacon.

Wang, M. (1997) The Rise of International Trade and Global Logistics in East Asia: The Case Study Emphasizing Asian NIEs, *Study of Shipping Economy*, 31, Tokyo: Japan Society of Shipping Economics, 117–35.

Wang, M. (1998) The Competitive Advantage of Northeast Asia Hub Ports – Kobe and Pusan: From the Viewpoint of Global Container Transportation, *The Bulletin of Japan Maritime Research Institute*, No. 379–381, Tokyo: Japan Maritime Research Institute, January–March.

Wang, M. (1999a) *The Knowledge of Global Logistics in East Asia*, Kyoto: Bunrigaku.

Wang, M. (1999b) The Rise of Asia and its Way towards the 21st Century, *Ritsumeikan Journal of Asia Pacific Studies*, 4, Kyoto: Ritsumeikan Center for Asia Pacific Studies Ritsumeikan University, 133–156.

Wang, M. (2000) *The Global Logistics System in East Asia*, Kyoto: Bunrigaku.

3
Network Economies of Global Alliances in Liner Shipping: The Case of Japanese Liner Shipping Companies

Shigeru Yoshida, Jian Hua Yang and Kwang Hee Kim

3.1 Introduction

International liner shipping experienced a great deal of innovation after the 1980s. This was caused by the development of Asian shipping to keep pace with economic development and the enactment of the revised US Shipping Act in 1984 as part of a tidal current of deregulation. As a result, international liner shipping, which was formerly dominated by developed countries, began to face big changes in market structure. Liner shipping conferences, with a history reaching back over a century and a half, began to collapse. Worldwide competition in the liner shipping market intensified with economic globalization and many traditional liner shipping companies went bankrupt. In such a situation, global alliances appeared that focused on the larger and more mature shipping companies and were different from the shipping conference, consortium or slot charter arrangements that had existed previously.

Although there has been important recent research on shipping alliances (for example, Ryoo and Thanopoulou, 1999; Brooks, 2000; Midoro and Pitto, 2000), until now it has not explicitly analysed the network economics on which this chapter especially focuses. Much research emphasizes the external network effect on the demand side. However, herein we also pay attention to the cost effects created by the economies of network scale on the supply side. Therefore, the aim of this study is to analyse statistically the network effects of alliance formation on both the demand and supply sides of Japanese liner shipping companies.

3.2 Importance of route networks on alliance formation

The characteristics of container shipping firms operating during the competitive era of the 1990s are indicated in various ways; an increase in competitors, an extension of competitive regions and areas of specialization, increased market entry and exit, and governmental deregulation. These features explain the globalization of both the liner shipping market and the enterprises that operate within it. Remarkably, the same tendency can be observed in the alliances of liner shipping.

The attributes of each alliance in international liner shipping are shown in Table 3.1. The common features of alliances that distinguish them from conventional consortium or slot charter arrangements can be summarized as the following:

- Partnership of famous shipping companies in different countries.
- Partnership on a worldwide scale.
- Partnership focusing mainly on the Asian market.
- Long-term and comprehensive partnership.

The liner shipping alliances have various aims that include: cost reduction by cooperation and the improvement of facilities; service enrichment in frequency and area served through the extension of capacity; and the mutual utilization of management resources. Within these aims, we focus on service enrichment from the viewpoint of network economics. In order to analyse the economics of the alliances, we shall examine how important the route network has been in the formation of the alliances.

Table 3.1 Attributes of each alliance in 2001

Alliance Name	Maersk/Sea Land	The New World Alliance (TNWA)	Grand Alliance	United Alliance	COSCO/K-Line/Yang ming
Members	Maersk Sea-Land	MOL Hyundai NOL(APL)	NYK P&O/Nedlloyd Hapag-Lloyd OOCL, MISC[2]	Hanjin DSR-Senator Choyang UASC	COSCO K-Line Yangming
Number of VSL	141	94	173	152	90
Capacity of VSL	692,450	390,000	742,800	567,200	347,000
TEU/ VSL	4,911	4,149	4,294	3,732	3,856
Start date	Jan. 1996[1]	Jan. 1998	Jan. 1998	Jan. 1997	Mar. 1998

Source: NYK (2001), 'The World Container Fleet and its Operations'.
Note: Number and capacity of vessels are estimated by NYK:
1. Maersk was acquired by Sea-Land in 1999.
2. MISC participates only in the Asia–Europe route.

Table 3.2   Number of routes of members in the Global Alliance in 1994

	APL	OOCL	Nedlloyd	MISC	MOL
Asia–N. America	5	5			6
Asia–S. America					10
Asia–S. Pacific					1
Asia–Africa			5		6
Asia–C. America					2
Asia–Australia		1	2	1	4
Intra-Asia	19	16	2	9	14
Australia–N. America			1		
Europe–Australia			1		
Europe–C. America			2		
Europe–Mid Asia			2		
Europe–Africa			4		
Europe–N. America (Atlantic)		6	4		
Europe–Pacific			1		
Europe–C. America			2		
Europe–Far East Asia		2	2	2	2
Intra-Europe			2		
Intra-America			5		
TOTAL	24	30	35	12	45

Source: *Containerization International Yearbook*, 1995.

First of all, we shall review the process of forming the Global Alliance. In the initial days following its formation, the Global Alliance comprised about 205 ships or 400,000TEU capacity and was the biggest amongst the alliance groups (Damas, 1995). Table 3.2 shows the number of routes operated by each member of the alliance before the formation of the alliance.

The following findings can be pointed out using this table and other resources; MOL served several routes mainly from Asia to America, Africa, Australia and Europe, and 14 routes in the intra-Asia area. MISC was competitive in the intra-Asia market, Nedlloyd was competitive in the Asia/Europe route as a mainline route (Anon, 1996a), OOCL mainly served the intra-Asia and Asia/North American routes, and APL mainly served and was competitive in the intra-Asian routes linked to both North America and feeder service routes (Anon, 1996b). After the alliance was formed, the group cooperated on the two big mainline trades, namely the Far East Asia/North America area having seven routes and the Far East Asia/Northwest Europe area having three routes. Thereafter, there arose several problems regarding the route network configuration of the entire alliance. These were as follows:

- The role of MISC in the alliance was merely as a feeder service company in the intra-Asian routes on which, before the alliance, MOL and Nedlloyd had insufficient services.

- In the case of OOCL, the company was not cooperating with, but was rather competitive with, MOL in the Asia/North America and intra-Asia routes.
- A similar circumstance appeared in the relationship between Nedlloyd and MOL in the Asia/Europe, Asia/Africa and Asia/Australia routes (Anon, 1999a).

Immediately after the formation of the alliance, intense competition between alliance groups occurred in the international liner shipping market and struggles or competitive disputes within a group occurred in order to acquire the initial power or superiority over the group. As a result, the Global Alliance had to be restructured as The New World Alliance (Anon, 1999b). It is, thus, justified to say that the history of the formation of the Global Alliance proves that it is very important for each alliance member to acquire a proper position with respect to their route network within the context of the entire network formation of the groups.

Next, we shall look into the Grand Alliance as a case of 'Alliances of complementary equals' (Bleeke and Ernest, 1995). Members at the time of the alliance formation were NYK, P&O, Hapag-Lloyd and NOL. Since NOL transferred to The New World Alliance after its merger with APL, and OOCL and MISC then joined as new members, at present this group consists of NYK, P&O and Nedlloyd, Hapag-Lloyd, OOCL and MISC. Table 3.3 shows the changes of partner of each alliance.

Table 3.3   Changes of partner of each alliance

Group name	Partners of the first generation (1995–1997)	Partners of the second generation (1998–2002)
Global Alliance ⤵ The New World Alliance	APL Nedlloyd MOL OOCL MISC	NOL/APL MOL HMM
Grand Alliance	Hapag-Lloyd NYK NOL P&O	Hapag-Lloyd NYK P&O/Nedlloyd OOCL MISC
Maersk/Sea-land	Maersk Sea-land	Maersk Sea-land
Hanjin/Tricon ⤵ United Alliance	Hanjin DSR-Senator UASC Choyang	Hanjin DSR-Senator UASC Choyang

A feature of the Grand Alliance has been to develop the network service evenly in both Europe and North American routes in comparison to The New World Alliance and the United Alliance. This was necessary because of the participation in the alliance of the big European companies, P&ONL and Hapag-Lloyd (Anon, 2000).

Inspection of the entire route network of the alliance reveals that NYK was most competitive on the Asia–North America route whilst P&ONL and Hapag-Lloyd were more competitive on the Asia–Europe route. Thus, by combining each member's route network, the alliance group was able to construct a worldwide network. Afterwards, NYK strengthened its Asia–Africa, Asia–Australia and intra-Asia routes, Hapag-Lloyd strengthened its Europe–Central America (or via Asia) and Europe–Mediterranean routes, and P&ONL strengthened each of the routes from Europe to Africa, the Mediterranean and Australia. In addition, OOCL expanded its China route and MISC did the same for the Asia–Australia route and the South East Asia region.

The strategic route development of the Grand Alliance to enrich the network service equally on both Asian and Europe routes shows which factors were important in the selection of alliance partners (Kadar, 1996). These factors were the nationality of the partners and the degree of complementarity of prospective partners. With regard to the complementarity issue, when each member tried to establish a new North–South route linked to an East–West mainline route (such as the Asia/Europe or trans-Pacific route), they selected new partners whose network routes were supplementary to each of their own and where true cooperation resulted. Therefore, such a partnership was naturally not limited to any country or region, and was not one aimed at acquiring a hub port, but rather to strengthen feeder services (Frankel, 1995; Robinson, 1998). The case in which the nationality of a partner becomes an issue is when alliance members try to approach and break into the market of new partners.

3.3   The network economics of a global alliance

Just as a global alliance is called a strategic alliance, an alliance in international liner shipping also exhibits features of competitive strategy. Strategic alliances in shipping specifically pursue economic benefits that are a little different from those which are sought in traditional freight rate and discriminatory competition. Though the alliance is characterized by direct economic benefits, that is, cost reduction by the joint use of many facilities and the coordination of the fleet, it also has an important indirect effect in terms of network economics, where there are two sources of economies (Katz and Shapiro, 1992). One is the external impact of the network scale of the alliances on the demand side, and the other is the economies of scope of the network relating to the quality of network service (i.e. high service frequency, wide service area, many service points and short transit time). The reason why network economies are considered is because liner services are offered as a network service. A liner service is basically a link between two ports, meaning that the shipping conference works on a route basis. In contrast, the alliance is changing the way in which characteristics of the network service are based on the route network. They have become more important with the increased globalization of the cargo owners. Therefore, it is necessary to grasp the effects of the alliance on service quality and the service network.

Improvement of service quality by an alliance

Because it is a network service, the improvement of liner shipping services involves attempting to increase the number of service points and service frequency levels. Although alliances have led to some expansion of container shipping services through an increase in the number of ports of call, it is an increase in the number of weekly service loops and service frequency that has been the major source of innovation. According to Table 3.4, the number of weekly service loops more than doubled from 23 to 52 on the North USA route and from 11 to 24 on the Europe route following alliance formation. While increasing the number of loops raises the service frequency to a specific port of call, another strategy has been to increase the number of ports of call. For example on the Asia–North American route as shown in Figure 3.1 We can observe that although each alliance had about 15 ports of call at the start of 1995, by the beginning of 2001 Maersk-Sea-Land had nearly doubled its number to 32 and the Kawasaki Kisen/Yang Ming/COSCO group (with quite a number of port calls in China and Southeast Asia) increased its number remarkably; by more than 40 per cent. Over the past five years, each alliance has increased the number of ports of call as service points quite considerably.

Table 3.4   Number of weekly service loops of each alliance by route in 2001

	Asia/North America	*Asia/Europe-Mediterranean*	*North America/Europe-Mediterranean*
Grand Alliance	6[1]	7[1]	3+(2)
The New World Alliance	9[1]	4	1+(2)
	7[1]	6	2
United Alliance	7[2]	5[2]	3+(4)
Maersk-Sea-Land	8	4	3
COSCO/K-Line/YMTC Evergreen/LT	6	4	2
Total	43	30	22

Source: Ocean Commerce Ltd., *International Transport Handbook;* NYK (2001) *The World Container Fleet and its Operations*
Note: Figures in [ ] are number of the services linked to other routes.
Figures in ( ) are number of the services cooperated by members.

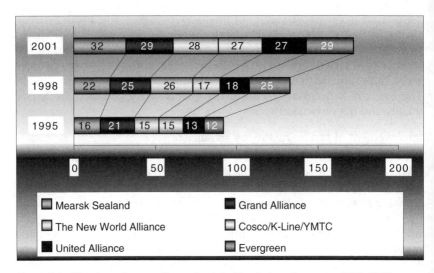

Figure 3.1 Number of port calls on the Asia–North America route, 1995–2001
Source: Ocean Commerce Ltd, *International Transport Handbook*.

The increase of service frequency and service points shows the importance of the network to liner services, to such an extent that it allows the
alliance to enable improvements to the quality of the service. The most
important point is that the improvement of service quality brings about
greater market share; an increase that is connected with the economies of
network scale. Generally, this has been illustrated by an S-shaped curve,
which specifies the statistical relationship between an increase in the share
of service frequency and the increase in market share that this brings
about. More specifically, this relationship has been found to also hold for a
network industry (Takahashi and Yoshida, 1995). This phenomenon is
explained by the existence of network externalities that are described in the
following section.

Network externalities and their effects on alliance market share

For the purposes of this chapter, a network externality is defined as being a
situation where the benefit from additional network scale surpasses the cost
of increasing the size of the network, and most notably, arises on the
demand side (Katz and Shapiro, 1985). The network externality that is
derived from alliance formation is to be found in the cargo owner's benefit
that it gives at little additional cost. This comes from the expansion of
network scale and improvement of network service quality that is implemented not by a single company, but by the alliance as a whole. The reason
why this sort of network externality (i.e. that is derived from alliance formation) is so important is that it yields continuous benefits, while the cost of

bringing these about is merely a relatively minor initial effort. The extension of the network scale through alliance formation, therefore, involves the trade-off of a low cost with a higher quality of service and can stimulate positive network externalities brought about by increasing the number of port calls, rationalizing the selection of ports of call, extending transportation capacity, shortening the navigation period and others.

There are various ways in which an alliance has an effect on market share. Immediately, the market share of a company increases up to the aggregate share of the group. The amount by which market share rises will depend upon the degree of trade-off between service frequency and size of ship deployed. In other words, it is possible that the increase in the sales value of the alliance is closely followed by the enlargement of ships deployed and an increase in service frequency. The enlargement of ship size has advanced remarkably following the advent of alliance formation, and the economies of scale in ship size has been greatly supported by increases in the value of sales. The alliance has enabled a company to enjoy the economies of ship size precisely because sales value at aggregate alliance level is much higher than at company level. Therefore, the optimum ship size based on the sales volume of the alliance is larger than that of a single company member. This has become apparent in the accelerating speed at which ship size has increased since alliances began to be established (Yoshida, 1999).

Table 3.5   Market share and service attributes of each alliance on the Asia/North American route

	Share of 1995 (%)	Share of 1998 (%)	ΔS (%)	Number of calling ports	Carrying capacity (TEU)	Average ship	Speed (Knots)
The New World Alliance	27.0	27.9	0.9	26	194,459	3,670	23.5
COSCO/ Yang Ming/ K-Line	18.5	19.1	0.6	17	106,440	3,226	21.5
Maersk/ Sea-Land	14.0	15.6	1.6	22	114,635	3,275	23.2
United Alliance	12.0	12.6	0.6	18	180,249	3,400	23.0
Grand Alliance	12.0	13.5	1.5	25	141,665	3,373	22.5

Source: Containerization International Yearbook and Conferences data.
Note: ΔS shows the difference in market shares between 1995 and 1998.
The other corporations are omitted in the table.
Except for market shares, each variable is observed in 1998.

Table 3.6   Correlation coefficients between changes in market share and other factors

	ΔS	Number of calling ports	Average ship size	Speed	Carrying capacity
ΔS	1				
Number of calling ports	0.613	1			
Ave. ship size	−0.131	0.663	1		
Speed	0.319	0.589	0.677	1	
Carrying capacity	−0.319	0.389	0.896*	0.669	1

Note: Figures are calculated from the data of Table 3.5.
Asterisk means statistical significance at the 5% level.

Using correlation analysis and on the basis of each alliance grouping, we will try to investigate what factors have influenced the change of market share (ΔS) on the North America route before and after alliance formation (1995 and 1998) (Yoshida, Yang and Kim, 2000). It is hypothesized that the factors which influence the market share are: freight rate or price, number of ports of call service frequency, speed of ship and carrying capacity; all of which are attributes of service quality. The results are as shown in Table 3.5.

From Table 3.6 we deduce that the highest correlation coefficient affecting the change in market share is the number of ports of call. The impacts of service frequency and ship speed seem to be of little significance because a weekly service has been adopted as the general basis for fleet operation, resulting in insignificant differences between each company in terms of ship speed. It is possible, therefore, to point out that the increase in the number of ports of call demonstrates the importance of network size or service points. In other words, it is clear that a network externality has been realized by the increase in the number of port calls (or service points) in liner shipping. Therefore, we state that the network is a strategic variable for all alliances.

3.4   Effects of alliances on Japanese liner shipping companies

Finally, we analyse statistically the alliance effect on both demand and supply sides, taking as examples three Japanese liner shipping companies, NYK, MOL and K-Line, each of which belong to different alliances.

Network effect on demand side: market share

As mentioned above, it is clear that network variables are very important for market share. We, thus, tried to observe the relation between the cargo volume carried by each of the three companies and network variables such as the number of ships, the number of loops, the number of port calls, the average ship size and the TEU capacity of each alliance. Japanese liner companies had been facing a situation in which their total market share at the global level had been continually in decline prior to alliance formation. The question as to whether the situation has changed or not is an interesting one. The cargo volumes shipped by the three companies are shown in Table 3.7. It can be clearly seen that only MOL has been influenced significantly by the alliance. The data indicates that NYK and K-Line cargo volumes showed very little effect due to the establishment of the alliances.

We then analysed the effect of alliances on the market share of the three Japanese companies. This was done by estimating market share functions with explanatory variables as follows: the relative unit cost of container movement, average ship size (as proxy variables of competitiveness), number of loops and port calls (as network variables). Notice here that the market share data are calculated on a domestic basis, that is each company's share of the total cargoes shipped by the three liner companies in percentages. The data used were panel data of the three companies and the

Table 3.7   Shipped cargo volumes in 1994 and 1999

	1994	*1999 (growth rate, %)*
NYK	16,285	16,961 (4.2)
MOL	14,638	17,151 (17.2)
K-Line	11,482	11,915 (3.8)

Source: Annual Reports and JSA Shipping Statistics.
Note: These data are calculated from freight revenues and others.
Units of shipped cargo volumes is 1,000 metric tons.

Table 3.8   Estimation results of market share functions

	(1)	*(2)*	*(3)*
Unit cost of container related	0.3284*		0.3227*
	(3.67)		(3.27)
Average ship size		–0.2541	–0.0280
		(–1.25)	(–0.152)
Number of loops	0.2028	0.1058	0.2098
	(1.36)	(0.57)	(1.32)
Number of calling ports	0.5075*	0.5309*	0.5080*
	(3.68)	(3.17)	(3.61)
Constant	–0.9009	1.4547	–0.7860
R–bar squared	0.697	0.551	0.684
Standard error	0.043	0.052	0.044
Variation of residual	0.042	0.06	0.043

Notes: Values within parentheses are t–values.
Asterisk means statistical significance at the 5% level.

observation period was from 1991 to 1999. The OLS method was utilized for statistical estimation.

The results of the estimation stage are shown in Table 3.8. Judging the results, the unit cost variable has the wrong sign from a theoretical point of view, the number of port calls was found to be statistically significant at the 5 per cent level and R-squared values were not low. It is clear that cost competitiveness as proxied by the average ship size and service point variables are effective in bringing about a change in the market share. It has already been mentioned that MOL recorded the highest increase in market share of the three companies. It is interesting to note that it is a member of The New World Alliance that records the largest number of port calls. In contrast to MOL, NYK as a member of the Grand Alliance has increased average ship size, but this has still led to its insignificance in the estimation of market share functions.

Although the question as to which alliance a company participates is clearly a critical one, the number of port calls could be the most important influential variable on market share and network effects.

Network effects on the supply side: cost effect

In general, the cost reductions resulting from alliance formation are based on cooperation in the deployment of facilities. However, whether or not network enlargement brings about this effect does not seem to have been investigated. We, therefore, analysed this effect by estimating the cost function including network variables. On estimation of the function, we used the 'Constant Economies of Scale' function together with the price of input. As a network industry, the transport sector exhibits various kinds of economy of scale, which are density, scope and network (Caves et al., 1985). The distinction between economies of scope and network depends on the purpose of the analysis. We, therefore, focus on the traditional economies of scale and network scale.

The cost analysed was the container-related cost excluding ship-related costs (capital and voyage cost) that were omitted due to a lack of available data. Observing the unit cost (per TEU) of the three companies, as shown in Figure 3.2, their trends are contrasting; that is, K-Line has continued to decline, MOL has lowered its costs remarkably following alliance formation, and, in contrast to MOL, NYK, has stagnated in recent years. It was only MOL that showed cost effects brought about by alliance formation.

We investigated statistically to what extent the difference of cost movements can be explained by the network variables of each alliance. Independent variables that influence the cost and that are included in the analysis are: route network variables (number of loops, number of ships, number of port calls), a competitiveness variable (average ship size), supply and demand variables (carrying capacity and shipped cargo tons) and the input prices of container and labour. The data type used in the analysis is

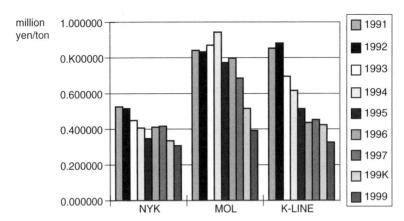

Figure 3.2   Changes in unit costs of three Japanese liner shipping companies, 1991–1999
Sources: Annual reports and JSA, *Shipping Statistics*.

Table 3.9   Estimation results of cost functions

Equation No.	(1)	(2)	(3)
Shipped cargo ton	0.6680*		0.3152
	(2.85)		(1.19)
Carrying capacity		0.6041*	0.4635*
		(3.64)	(2.29)
Number of calling ports	0.1018	0.1714	0.0423
	(0.53)	(1.21)	(0.24)
Input price (container)	0.9737*	0.7806*	0.7012*
	(2.47)	(8.46)	(6.20)
Input price (labor)	0.5178*	0.8098*	0.8471*
	(5.94)	(2.20)	(2.32)
Constant	–5.7321	–6.0302	–6.3702
R–bar squared	0.7390	0.7770	0.7812
Standard error	0.0463	0.0431	0.0427
Variation of residual	0.0021	0.0019	0.0018

Notes: Values within parentheses are t–values
*indicates statistically significant at the 5% level.

panel data of the three companies from 1991 to 1999. Estimation is done in the logarithm using the OLS method.

Based on the sign of the parameters, the coefficient and t-value of the estimated equations, the results of the estimation of the cost functions are shown in Table 3.9. Many equations were estimated and the final significant equations consist of the shipped cargo tons, carrying capacity, number of port calls (or loops), and input price variables. The other variables were rejected because of the low significance levels of the t-value and the meaningless sign of the parameters.

From the results of the estimation, we can point to the existence of scale economies in carrying capacity and shipped cargo tons because the values of the parameters were smaller than one. Also, the number of port calls variable is not significant at the 5 per cent level. However, based on equation (2) having a statistical problem, network economies have to exist. Therefore, we deduce that network variables are effective in reducing costs through the expansion of service network.

3.5   Conclusions

In conclusion, the factual findings that have been deduced from the analysis contained in this chapter, are as follows. First, it is the route network of the members that plays the most important role when the composition of the alliances was decided. Secondly, it is the number of port calls as service points that have the largest effect on the network externality of the alliances. This implies that the extension of this factor to correspond to customers' needs has made a remarkable difference among the alliances. In other words, the other service attributes are of no significance in the alliances. Thirdly, the network scale shown by the number of port calls is effective in influencing changes in the market shares of Japanese liner shipping companies. Finally, regarding the economics of network scale, although the statistical estimation was not significant, we can say that Japanese liner shipping companies have achieved cost reductions through the network extension brought about by the alliances.

The alliances, which appeared in international liner shipping after the collapse of the conference, corresponds with the innovation of the international liner shipping market and international economic globalization, for which strategic characteristics and network economies are key concepts. In the future, it is unknown whether the alliance will bring about a new order in the liner market or will replace the traditional shipping conference. This is because the essence of various alliances is strategic and any strategy is dependent upon the commercial environment in which the industry operates.

References

Anon (1996a) Growth through Diversity at MISC, *Lloyd's Shipping Economist*, 18(8), 23–6.

Anon (1996b) APL Adopts a Broad Global Perspective, *Lloyd's Shipping Economist*, 18(11), 23–6.

Anon (1999a) Three's a Crowd, *Lloyd's Shipping Economist*, 21(3), 12–14.

Anon (1999b) Shipping Seeks New Global Role, *Lloyd's Shipping Economist*, 21(11), 7–10.

Anon (2000) Consolidation in Container Trades, *Lloyd's Shipping Economist,* 22(2), 21–4.

Bleeke, J. and Ernest, D. (1995) Is Your Strategic Alliance Really a Sale?, *Harvard Business Review,* Jan.–Feb., 97–105.

Brooks, M.R. (2000) *Sea Change in Liner Shipping Regulation and Managerial Decision-Making in a Global Industry,* Amsterdam: Elsevier.

Caves, D.W., Christensen, L.R., Trethway, M.W. and Windle, R.J. (1985) Network Effects and the Measurement of Returns to Scale and Density for U.S. Railroads, in A.F. Daugherty (ed.), *Analytical Studies in Transport Economics,* Cambridge: Cambridge University Press.

Damas, P. (1995) The Global Count: 4+4+2, Containerization *International,* July, 59–60.

Frankel, E.G. (1995) Economic Incentives for Liner Alliance and Impact on Container Feeder Shipping, *Proceedings of IAME Conference,* Cambridge, MA: MIT.

Japan Ministry of Transport (1999) *White Paper on Japan Shipping,* Tokyo: Japan Ministry of Transport.

Kadar, M. (1996) The Future of Global Strategic Alliances, *Containerization International,* April, 81–5.

Katz, M.L. and Shapiro, C. (1985) Network Externalities Competition and Compatibility, *American Economic Review,* 75(3), 424–40.

Katz, M.L. and Shapiro, C. (1992) Product Introduction with Network Externalities, *The Journal of Industrial Economics,* 40(1), 55–84.

Midoro, R., and Pitto, A. (2000) A Critical Evaluation of Strategic Alliances in Liner Shipping, *Maritime Policy and Management,* 27(1), 31–40.

Robinson, R. (1998) Asian Hub/Feeder Nets: the Dynamics of Restructuring, *Maritime Policy and Management,* 25(1), 21–40.

Ryoo, D.K. and Thanopoulou, H. (1999) Liner Alliances in the Globalization Era: A Strategic Tool for Asian Container Carriers, *Maritime Policy and Management,* 26(4), 349–67.

Takahashi, N. and Yoshida, S. (1995) *International Transport,* Tokyo: Sekai Shiso Co.

Yoshida, S. (1999) Competitive Formation of Japan Shipping and its Change, *Study of Maritime Transportation,* 48, Oct.

Yoshida, S., Yang, J.H. and Kim, K.H. (2000) The Global Alliance in the International Liner Shipping, *Proceedings of TECHNO OCEAN,* 1, September.

4
Choice of Financing and Governance Structures in the Transport Industry: Theory and Practice

Stephen X.H. Gong, Michael Firth and Kevin Cullinane

4.1 Introduction

The purpose of this chapter is to develop a descriptive theory of corporate finance and corporate governance for the world transport industry. More specifically, we examine the economic, institutional and industrial factors that lie behind a transportation company's choice of different financing mechanisms and governance structures (collectively called 'financing methods'). This combined treatment of corporate finance and corporate governance is in line with Williamson (1988), who argues that 'the supply of a good or service and its governance need to be examined simultaneously'. This is a step further than the traditional approach in corporate finance, which tends to limit its investigation of a firm's choice of financial structure (i.e. debt versus equity) *in aggregate*, with no follow-on inquiry into the resultant governance structure at an individual firm or industry level. The value of the level of aggregation and abstraction in neoclassical financial theory, along with the simplifying assumptions that are usually made, lies in their potential for clarifying difficult analytic issues, but it also rules out important practical considerations which may influence financial system form and functions (Neave, 1991).

In this study, we draw on insights from institutional economics (of which transaction cost economics is a fast-growing branch) and development economics and explain a firm's financing methods within the context of a wider socioeconomic environment and under alternative behavioural assumptions. The result of this analysis is an eclectic theory of transactional factors that help to shape the form and function, as well as their change over time, of the financing methods used in the world transport industry. In combination with insights from traditional finance theory, this has the potential to provide a better explanation of the industry's financing patterns as observed in practice.

Section 4.1 compares the neoclassical (modern) theory of finance with the transaction cost economics approach to corporate finance and corporate governance. This also provides the justification for our industry-specific study of financing behaviour. Section 4.2 reviews some major theories and empirical studies of firms' financing choices. Next, section 4.3 has a discussion of the relevance and applicability of these theories to the transport industry. The economic, institutional and industrial characteristics of the transport industry are considered in section 4.4, leading to the formulation of a descriptive theory of corporate finance and corporate governance. Finally, section 4.5 concludes by identifying some implications and directions for future research.

4.2 Neoclassical finance theory versus transaction cost economics versus practical observation

In his article introducing the idea of transaction cost economics to corporate finance and corporate governance, Williamson (1988) notes that neoclassical finance theory and transaction cost economics (TCE) are mainly complementary approaches to economic organization; both have helped and will continue to inform our understanding of different forms and functions of organization. Agency theory (Jensen and Meckling, 1976), a landmark development of finance theory, and transaction cost economics both come in different forms, but each approach is concerned (either directly or by implication) with different types of economic organization and their effects on efficient contracting and control. Agency theory (AT) is concerned with the separation of ownership from control, while TCE traces its origins to vertical integration and is subsequently applied to the governance of contractual relations. The key commonalities and differences between the two include:[1]

- Managerial discretion and efficient contracting: Both TCE and AT take exception with the traditional theory of the firm whereby the firm is regarded as a production function to which a profit-maximization objective has been ascribed. Instead, TCE regards the firm as a governance structure and AT considers it a nexus of contracts. Both rely on substantially identical behavioural assumptions, i.e. that human agents are subject to bounded rationality (leading to incomplete contracting) and are given to opportunism (moral hazard and agency costs in AT). This is the basis of the efficient-contracting orientation to economic organization that is common to both approaches. AT examines contracting mainly from an *ex ante* incentive-alignment point of view, while TCE is more concerned with crafting *ex post* governance structures.
- Basic unit of analysis and agency/transactions costs: The most important difference between TCE and AT lies in the choice of the basic unit

of analysis. TCE regards the transaction as the basic unit of analysis, while AT focuses on the individual agent. Both are micro-analytic in nature and both imply the study of contracting. Whereas identifying the transaction as the basic unit of analysis leads TCE naturally to an examination of the principal dimensions with respect to which transactions differ, use of the individual agent as the elementary unit of analysis has given rise to no similar follow-on effort in AT. In TCE, asset specificity (including human capital specificity) is one of the key dimensions with respect to which transactions differ, and one resulting refutable implication is the proposition that agents align transactions (which differ in their attributes) with governance structures (the costs and competences of which differ) in a transaction-cost economizing way. Assessing the comparative efficacy of alternative governance structures for harmonizing *ex post* contractual relations is the distinctive focus and contribution of TCE. In AT, the existence of agency costs (monitoring expenditures of the principal, bonding expenditures of the agent, and the residual loss), which are assumed evident to prospective transacting parties, means that equity and debt will be priced on the basis of the projected performance of the firm after agency costs have been taken into account. In other words, the entrepreneur will bear (*ex ante*) the entire wealth effects of these expected costs so long as the market anticipates these effects.

- Debt and equity as financial instruments versus governance structures: TCE's focus on individual transactions and the implied imperative to align transactions with governance structures in a transaction-cost-minimizing manner leads to the identification of attributes of different transactions and the resulting contractual arrangement that is chosen to match these attributes. Whereas most prior studies of corporate finance treat debt and equity as merely financial instruments, TCE regards them as different governance structures that are chosen in a discriminating way.

Apart from the influence of agency theory, modern corporate finance has been revolutionized by the famous Modigliani and Miller (1958) (M&M) capital structure irrelevance propositions. Basically, these propositions state that, under certain assumptions (see discussion below), a firm's value is determined by its real assets/investments and not by the type of instruments used to finance the investments. The key contribution of the M&M propositions is that they provide the basis for rational investment decision making within the firm; and by showing what does not matter, they also show (by implication) what does (Miller, 1988). Since many of the necessary conditions for the M&M capital irrelevance propositions are violated in real-life situations, it is important for the financial manager to know what kinds of market imperfections to look for (Brealey and Myers, 2000).

Being practically oriented, this study aims to examine what factors, in a real-life marketplace, shape or influence the forms and functions of financing that a firm adopts, and how these vary across countries (with different socioeconomic and institutional characteristics) and evolve over time. This industry-specific, micro-analytic approach contrasts with that found in traditional finance, which tends to abstract from industry- or firm-specific characteristics and country effects by looking into firms' financing decisions in aggregate and under various simplifying assumptions. The value of the latter approach is obvious (one of which being that it is tractable and its inferences are probably generalizable), but since the value of a model depends on the use to which it is put, and given our aim to explore the factors influencing the financing of transportation companies in real life (where one or more of the restricting conditions are unlikely to hold), it is only natural that we do not restrict ourselves to the M&M propositions or its counterparts in modern finance. At this juncture, it should be noted that the view that capital structure is literally irrelevant in corporate finance is actually far from anything M&M ever actually said about the real world applications of their theoretical propositions (Miller, 1988). In choosing to look to reality to explain the financing decisions of industry firms, we actually follow M&M's own advice to relax their assumptions in the direction of greater realism and relevance in an industry-specific context.

Many authors have noted the merits of an industry-focused study. In discussing the testing of economic hypotheses using homogeneous groups, Elton and Gruber (1970) show that pooling data for all firms (with distinct characteristics) without controlling for latent variables can destroy the true underlying relationship among variables. The identification of homogeneous groups of firms (which helps to hold constant some latent variables across firms) can form a more meaningful basis for the testing of hypotheses in the field of finance (Elton and Gruber, 1970: 587). In their highly acclaimed study of the cost of capital, Miller and Modigliani (1966) use the utility industry as the testing ground for techniques of estimating the cost of capital because: 'They permit us to have both a large sample and one in which the component firms are remarkably homogeneous in terms of product, technology, and market conditions' (p. 334).[2]

Neoclassical (modern) financial economics is essentially non-institutional, but institutions matter in ways not hitherto acknowledged or even imagined (Williamson, 1996: 192). This is also the stance taken by John Lintner, one of the creators of modern financial theory, who 'believed that the institutional setting of the financial markets, including legal and regulatory restrictions as well as elements of business organization and practice, can and does importantly affect financial behaviour and the resulting market outcomes' (Friedman, 1985: ix). Rybczynski (1984) also notes that the way an industry is financed must be considered against the background of the evolution of financial systems and the way the process of economic development has preceded in the past and is preceding at present.

In the sections that follow, we develop a theory of corporate finance and corporate governance in the world transport industry, based on a combination of financial economics, institutional economics, development economics, and observations of current practice. More specifically, we consider the following questions: What is the actual experience of using different financial instruments and governance structures by the transport industry over the last several decades? Are there cross-country or intra-industry differences in the industry's financing methods? How does the product market environment and/or the information environment that firms face influence their corporate financing decisions? How important are institutional factors like government and commercial organization involvement (e.g. subsidies, shipyard finance, taxes) in shaping firms' financing choices? Do specific firms have individually optimal financial structures? Does the entire industry (or a sub-sector thereof) have an optimal aggregate financial structure? If so, what are the determinants?

These are essentially the same questions asked by Friedman (1985), and it is beyond the scope of this current study to provide concrete answers to any of them, let alone all of them. Nevertheless, we hope that by asking the right questions, we can stimulate sufficient interest among future researchers who, through a series of coordinated studies, may subsequently shed light on these important questions that remain hitherto unanswered.

4.3   Review of the literature on corporate finance

It is usually the position in modern finance not only to separate the financing decision from the investment decision, but also to look at the mix of different financing instruments (i.e. the capital structure) as a 'fundamentally marketing problem' or as an outgrowth of information asymmetry (Myers and Majluf, 1984). Epitomizing the capital structure argument is the famous Modigliani and Miller (1958) proposition that 'financing decisions do not matter in perfect markets' (hereinafter, M&M I).[3] Recognizing the existence of market imperfections in reality (e.g. tax), finance theorists have proposed alternative explanations for the observed discrepancy between the M&M I proposition and real-life practice (e.g. Modigliani and Miller, 1963; Miller, 1976, 1988; Modigliani, 1982, 1988). Nevertheless, such alternative theories still fall short of generating testable hypotheses/predictions consistent with real-life observations, in particular in industry-specific settings. For a comprehensive survey of these theories and related empirical studies, see Allen and Winton (1995), Swoboda and Zechner (1995) and Maksimovic (1995). All of these authors, in their conclusions, point out that there remain many unresolved issues with regard to the financing decisions of firms. In this section we focus on three 'mainstream' theories. The first is the classic Modigliani and Miller (1958) capital structure irrelevance proposition (without tax). The second is the 'pecking order theory' proposed by Donaldson (1961) and popularized by Myers and Majluf (1984) and Myers (1984). The third is Williamson's (1988) transaction cost economics approach to a combined treatment of corporate finance and corporate governance. The main arguments of these three theories and some representative empirical studies are summarized below, while their relevance and applicability to the transport industry are discussed in a subsequent section. With the exception of Williamson (1988), it should be noted, these theories of corporate financing do not address the issue of the governance structure that follows a particular financing method. This latter aspect of corporate financing is a parallel focus of our chapter.

M&M's capital structure irrelevance propositions

Modigliani and Miller (1958) show that in perfect capital markets (without tax[4]), a firm's value is determined by its real assets, not by the securities it issues and, as a result, capital structure (the firm's mix of different securities) is irrelevant as long as the firm's investment decisions are taken as given. The arbitrage-based proof offered by Modigliani and Miller (1958) – see Brealey and Myers (2000) and Welch (1995) for an example – largely hinges on the presumption that both the levered firm and the unlevered firm generate exactly the same net operating income (agreed by and known to all) and that the capital markets are perfect and well-functioning (with no transaction costs). Given these and other assumptions, it is not difficult to show that the two firms have the same overall market value of debt and equity, no matter how they are packaged. Thus, it may be fair to say that the M&M conclusion resides in its presumptions. The important point in judging the value of a theory, of course, is to see whether its implications or predictions are borne out by real-life observations (Friedman, 1953).

It is obvious that in real life, corporate financing behaviour is not consistent with M&M's theorem. For example, while M&M predicts that capital structure does not matter in perfect capital markets, there exist clear patterns in the financial structure of specific industries (see Long and Malitz, 1985; Brealey and Myers, 2000). This discrepancy may be due to one or more of M&M's conditions not obtaining in practice, and it is useful to identify which of these conditions are violated in real life. Existing theories may be refined as a result.

In summarizing M&M I, Brealey and Myers (2000) remark that: 'We believe that in practice capital structure *does* matter... If you don't fully understand the conditions under which MM's theory holds, you won't fully understand why one capital structure is better than another. The financial manager needs to know what kinds of market imperfection to look for' (p. 474).

The most serious market imperfections are often those created by government. Examples include differential tax treatment (between debt and equity, or on the corporate level versus the personal income level), government grants/subsidies for some types of strategic investments (e.g. Title XI in the USA and Keihek Zoseon in Korea; see Lee, 1996), and other institutional factors. To understand the reason why transport financing takes on the patterns that we observe, it is undoubtedly important that we take into account not only the financial environment, but also the non-financial environment.

The pecking order theory

Myers (1984) contrasts the static trade-off hypothesis with the pecking order theory of financial structure. The static trade-off hypothesis posits that a firm's optimal financial structure (which varies from firm to firm) is determined by a trade-off of the costs and benefits of borrowing, given that the firm's assets and investment plans are held constant. The benefits of borrowing mainly come from the interest tax shield, and the costs from potential financial distress. Firms with safe, tangible assets and plenty of income to shield ought to have higher debt ratios than unprofitable firms with risky and intangible assets. Although these predictions are broadly consistent with observation, according to Brealey and Myers (1991), the static trade-off theory cannot explain why some of the most successful companies tend to have the least debt.[5] In this respect, the pecking order theory seems to do a better job. The pecking order theory of capital structure states that firms prefer internal finance and adapt target dividend payout ratios to their investment opportunities. When internal funds are insufficient for investments and firms have to use external finance, they will issue the safest securities first, for example debt, convertible bonds and, as a last resort, equity.

Baskin (1989) conducts an empirical test of the pecking order theory. Focusing on 378 large firms from the 1960 *Fortune 500* which had data available on Compustat in 1984, he finds that more profitable firms tend to have less (book value) debt, and he interprets this result as consistent with the predictions of the pecking order theory (and contradicting the static theory of optimal capital structure). Needless to say, this result is also consistent with other (perhaps more ad hoc) theories.

While the pecking order theory is consistent with implications of information asymmetries and the agency theory of Jensen and Meckling (1976) and other well-established facts (e.g. higher costs and more stringent market discipline for the use of public equity), several difficulties present themselves in the testing of the pecking order theory. Firstly, it is difficult to establish a line of demarcation by which the theory is to be accepted or rejected – the choice seems quite arbitrary. It can be quickly accepted or rejected, depending on one's criteria. Secondly, some of the results of many empirical studies can be shown to be consistent with both the pecking order theory and alternative theories. In short, both the static trade-off theory and the pecking order theory are successful in explaining some observed differences in the capital structures of firms, but are less successful in explaining others (Brealey and Myers, 1991). Therefore, despite the merits of both hypotheses, neither seems to be the whole truth.

The transaction cost economics (TCE) approach to corporate finance

Williamson's (1988) TCE approach to corporate finance starts by assuming only two forms of finance – projects must be financed entirely by debt or by equity but not both. He also assumes that projects can be arrayed from least to most in terms of their asset specificity, i.e. the extent to which the asset in question may be redeployed for alternative purposes without any substantial loss of value. Furthermore, it is assumed that debt is a governance structure that works almost entirely out of rules, which imposes a number of restrictions on the debtor: regular debt repayments, conforming to pre-specified liquidity tests, pre-emptive claims of the financier in case of default against the assets financed, etc. Different debt-holders will realize different recovery of value in the degree to which the assets in question are redeployable.

Since the value of a pre-emptive claim declines as the degree of asset specificity deepens, the terms of debt financing will be adjusted adversely against the debtor. Faced with increasingly adverse terms as asset specificity increases, the debtor might have to sacrifice some of the specialized investment features in favour of greater redeployability. But the more adverse financing terms associated with assets of greater specificity may be avoided by inventing a new governance structure (call it equity) to which suppliers of finance would attach more confidence, and this may be achieved by giving the equity providers closer control of the company, for example, through their access to internal information, their decision-making power through members of the board of directors, and residual claims on the firm's earnings and liquidated assets throughout the firm's life. Therefore, equity evolves as a governance structure to reduce the cost of capital for projects that involve limited redeployability (greater specificity) and to preserve value-enhancing investments in specific assets. It can thus be seen that debt and equity are actually governance structures (rather than merely financial instruments representing differential claims on firm value) that are created to match investment/asset characteristics with the costs of organizing such investment activities.

Thus, according to Williamson, a transaction-cost-minimizing reasoning supports the use of debt or rule-based financing for redeployable assets, while non-redeployable assets are financed by equity or discretion-based financing methods. This clearly has refutable implications as to whether there is a systematic pattern in a firm's financing of projects along the dimensions of asset specificity and degree of uncertainty.[6]

Motivated by the transaction cost economics paradigms attributable to Williamson (1979), Klein et al. (1978), and Monteverde and Teece (1982), Palay (1984) studies the factors that lead corporations to adopt particular forms of governance structure in rail freight contracts, including forms of contract and contractual terms. By cross-classifying rail freight contracts based on their terms and conditions against characteristics of investments (e.g. type and redeployability of equipment, transaction frequency), Palay (1984) provides evidence in support of the basic TCE proposition that, as investment characteristics become more transaction-specific, the associated governance structure becomes increasingly unique to the parties and transactions it supports.

In another application of TCE to the US airline and airport industry, Langner (1995) examines the relationship between contractual aspects (e.g. contract duration) of slot allocation and transaction-specific characteristics such as asset specificity and the frequency of transacting. His analyses using historical and current practices of transacting in slots in the United States indicate that the governance structures (i.e. contracting arrangements) between airlines and airports are related to specifics of the transacting parties, the frequency of transacting, as well as asset (including human asset) specificity, a finding that is broadly consistent with predictions of the TCE.

It is obvious from the above discussions that the corporate financing theories we have reviewed are not strictly opposed. For example, both the static trade-off theory and the TCE suggest that the form of financing is related to characteristics of the asset being financed. Also, both the pecking order theory and the TCE approach predict that equity is the financial instrument of last resort, albeit for different reasons.

As noted in an earlier section, one of the key differences among these theories is the unit of analysis: For transaction cost economics, the transaction is the basic unit of analysis and this focus on the transaction naturally leads to an examination of the principal dimensions with respect to which transactions differ. In contrast, the modern finance approach uses the individual agent as the elementary unit of analysis, and this focus on the agent (firm) leads to the study of financial structure choices on the level of the firm (distinct from its activities) but does not give rise to follow-on efforts to investigate the implications of the financing activity on the governance structure adopted (Williamson, 1988). Compared with the modern finance approach, the transaction-cost economics approach seems able to generate a wider class of testable implications that are applicable not only to corporate finance, but also to other economic organization issues. The wide spectrum of applications of transaction cost economics (among them corporate finance and corporate governance), including some empirical studies, is summarized in Williamson (1996), Williamson and Masten (1999) and Shelanski and Klein (1995).

The test of a theory is whether it yields meaningful predictions that are consistent with real-life observations. It is not the intention of this study to test the comparative validity of any of the existing theories of corporate financing behaviour. Instead, we aim to combine insights from these (and other) theories and develop a descriptive theory of corporate finance in the transport industry context. In the next section, we discuss the relevance and applicability of the pecking order theory and transaction cost economics to corporate financing behaviour in the world transport industry.

4.4 Relevance and applicability of existing corporate financing theories to the transport industry

Some important questions we attempt to answer in the study of financing behaviour in the transport industry are: (i) What determines whether a transportation company will choose between private debt (e.g. bank loans) and public debt (e.g. corporate bonds), or between internal equity and external equity, or some hybrid instruments (e.g. leasing)? (ii) What is the economic or institutional rationale behind the financing choices observed in practice?

Although there exist no systematic studies of the relative importance of different financing methods in the transport industry, there does exist a body of anecdotal evidence, which suggests that a bank (term) loan is the most favoured method of financing transport investments, in particular shipping investments. Stopford (1997), for example, remarks that commercial bank loans (where banks often provide up to 70 per cent of the total funds) have been by far the most popular form of finance for ships. He also notes that the type of finance available to the shipping industry has gone through distinct phases, and that such changing patterns of financing over time are related to the industry's own character (e.g. the shipping cycle, changes in the financial community's perception of risk and return in shipping, and institutional factors – see also Stokes, 1992, 1996). McConville and Leggate (1999), while concurring with Stopford and others, also note that the issuance of public bonds in the shipping industry (mainly in the United States and Western Europe) has been quite limited in number (34 as of 1999, as compared to over 300 public equity issues) and has been a comparatively recent phenomenon (with 70 per cent only in issue since late 1997). Thus, the comparative popularity of public equity over public debt seems to contradict the prediction of the pecking order theory – but see below for an explanation of the predominance of private (bank) debt over all other financing methods, a phenomenon that somewhat reverses this conclusion.[7] Elsewhere, Cullinane and Gong (2002) provide evidence that, at least as far as the Chinese mainland is concerned, evolutions in the country's economic and financial systems are responsible for the increasingly popularity of using public equity for financing transport investments in China, a nation that had previously relied primarily on state financing. However, no attempt has been made in any existing study of transport finance to offer any economic rationale for the observed financing patterns, their variations among countries, and their changes over time. This is an area to which this current study aims to make an original contribution.

Among the three theories outlined above, it appears that only the pecking order theory and transaction cost economics are able to generate empirically testable predictions and hypotheses regarding the questions raised earlier. Brealey and Myers (2000), for example, suggest the following testable hypotheses in the case of the pecking order theory:

- Firms prefer internal finance to external finance.
- If external finance is required, firms issue the safest security first. That is, they start with debt, then possibly hybrid securities such as convertible bonds, and then equity as a last resort.

Because equity ranks both on the top of the list (via retained earnings) and at the bottom, the pecking order theory does not presume a preference of debt over equity or vice versa (Brealey and Myers, 2000). As to the main reasons why firms prefer internal finance to external finance, Donaldson (1961), Myers (1984) and Brealey and Myers (1991) point to managers' reluctance to face the glare of publicity (in a stock issue), the restrictions imposed by lenders (in a bank loan), and the high issue cost of raising public equity. However, the pecking order theory does not tell us whether and why public debt or private debt will be used as a first priority.[8]

The pecking order theory, moreover, does not suggest what forms of governance structure (e.g. contract terms) will be adopted in various financing activities (for example, when debt is used). Similarly, the pecking order hypothesis cannot explain why financing methods in the same industry may differ across countries and vary over time. In comparison with the modern finance approach, under the New Institutional Economics[9] (NIE, of which transaction cost economics is a part), the transaction is made the unit of analysis, and the critical dimensions along which transactions differ are identified, including frequency of occurrence, uncertainty, and asset specificity (Williamson, 1996: 105). Since these factors vary from transaction to transaction, and from firm to firm, as well as over time, it predicts no *ex ante* optimal capital structure or financing choice that is applicable to every firm or that remains unchanged over time. Instead, the NIE/TCE approach to economic organization leads to predictions of financing choice only at a transactional level; it also gives rise to predictions about the governance structure (e.g. contractual arrangements) most likely to be used when a particular financing choice is indeed made. The NIE/TCE approach is also more eclectic, since it studies the financing choice behaviour of firms within a wider framework of the economic development (including the financial system), and transactional attributes (e.g. characteristics of the industry and the firm, including both the financier/lender and the fund-seeker/borrower).

Therefore, judging from the range of implications/predictions that the theory encompasses and their conformance with observations, it seems that the NIE/TCE approach is more relevant to our investigation. In the section that follows, we propose a descriptive theory of financing behaviour in the transport industry based mainly on the New Institutional Economics and drawing on insights from financial economics, development economics, and practical observations. This culminates in the formulation of a testable proposition to be empirically examined in future studies.

4.5 Corporate finance and corporate governance in the transport industry: a descriptive theory

To construct a theory of the choice of financing methods (as in any economic theory of choice), one has to start with the question of classification (Fama and Miller, 1972: 4). Following the general economic theory of choice, we divide the many separate factors bearing on any choice into two classes: The first class is the 'opportunity set' or the 'constraint set', and the second the decision maker's 'tastes' or preferences. In financial economics, the first class of factors influencing a firm's choice of financing is usually taken as given (exogenously determined) and the subject's preferences/ expectations are often assumed to be homogeneous. While this is in line with the traditional approach in positive economics, it overlooks an inherently crucial determinant of financing choice behaviour as observed in reality, i.e. the circumstantial factors (for example, the opportunity set of available financing choices) to which firms are subjected at various times. Financial economics is essentially non-institutional, but institutions matter in ways not hitherto acknowledged or even imagined (Williamson, 1996). According to Williamson (1993), a contractual approach to the study of economic organization informs the analysis of corporate finance and corporate governance quite generally, and cross-country studies need to be informed by significant legal (political and institutional) differences. Once the wider economic environment is taken into account, the evolutionary nature of the financing and governance mechanism is also predictable since financial markets/systems themselves change over time (Rybczynski, 1984; also see Taggart, 1985 for US evidence).

The second class of factors, that of the decision maker's (fund-raising firm's) preferences (idiosyncrasies), is also given greater attention in this study than in the finance literature. Modern financial theory is built on the assumptions of no (or inconsequential) transaction costs and well-specified (perfect) information conditions. This category of theory further assumes that financial agents have the same well-specified probability distributions (without showing how such are determined), and formulate and solve complex decision problems precisely and costlessly, even in the face of uncertainty and market imperfections. Given such hyper-rationality, agents are supposed to clearly understand the relations between alternative decisions and the impact of their decisions on their respective wealth positions. Therefore, the behaviour of these agents will be wholly rational, utility maximizing and predictable. Investors, on the other hand, are also assumed to be rational, utility-maximizing agents, able to see through any massaging of financial information. Hence, capital structure does not matter; neither do dividends.

In reality, the choice of an optimal capital structure for a given firm is a subjective decision. Agents (entrepreneurs, investors and bureaucrats alike) have bounded rationality (Simon, 1957), and there are expensive information costs, transaction costs, cost of financial distress and agency costs involved in decision making. These are the inevitable consequence of, and are exacerbated by, the existence of uncertainty/risks. Given such limitations, neoclassical financial theory is better at explaining well-established practices, but is poor in explaining creative or exploratory aspects of finance, the dynamics of technological changes, or the evolution of sophisticated financial systems from primitive ones. The reason is that financial system change largely occurs under uncertainty rather than objectively specified risk, and many of the interesting features of uncertainty are assumed away when decision problems are formulated in a manner conducive to neoclassical investigation (Neave, 1991). In comparison, the New Institutional Economics/TCE is well suited for the purpose of investigating choice behaviour under complex situations, since it takes into account current practices and other real-life imperfections in developing a theory to explain financial system organization and change.

In making a recommendation to implement research on cross-country studies from a transaction-cost economics perspective, Williamson (1993) proposes that country studies can usefully benefit from taking measures on, and making comparisons and assessments with respect to, some features of the institutional environment and institutions of governance. One of these features is the analysis of assets: 'An inventory of human and physical assets with reference to their industry- and firm-specific qualities is needed to evaluate economic organization' (Williamson, 1993: 53).

The TCE imperative: align financing structure capability/costs with transaction requirements

According to TCE, a financing mechanism is matched with a particular transaction on the basis of cost-effectiveness, capability, and the transaction's particular requirements. With regard to demand for (supply of) financing, this theory predicts that firms seeking funds (financiers) will choose between available alternative financing and governance mechanisms on the basis of perceived *overall* costs and benefits. Each choice is made on a transaction-by-transaction basis, aligning each transaction's requirements with governance capabilities and costs (hence many different kinds of alignments will coexist in any financial system at any given time; see Neave, 1991). Different sets of finance and governance structures are selected according to the agents' goals, the way they attempt to achieve these goals, the information conditions under which they act, and the costs and technologies of transacting (Neave, 1991).

When the transaction is made the basic unit of analysis, the most important dimension of which is asset specificity, TCE predicts that finance and governance structures will vary from transaction to transaction, even for the same firm. More specifically, projects for which physical asset specificity (redeployability) is low to moderate *ought to* be financed by debt (and, correspondingly, contractual terms suitable for the given asset specificity are used). As asset specificity becomes greater, however, the costs (including interest payments, financial distress costs, and covenant restrictions) of debt financing increases relative to the benefits, and now equity should be used instead. The influence of asset type or specificity on transport financing is well reflected in comments by Macquarie Corporate Finance Limited (2000: 41):

> Leasing is perhaps the most common form of finance in the aviation sector with virtually all of the world's airlines undertaking some form of lease financing...The rail sector has traditionally not used leasing as a means of finance because it has been dominated by government-owned organizations (private sector lease financing has been unable to match government's cost of funds) and because of the specialized and often non-standardized nature of the equipment. As the sector becomes more open to private operators through privatization and competition policy, a greater usage of leasing seems likely.

In the context of transport financing, the New Institutional Economics/ TCE theory also implies that financing and governance mechanisms (the form of financing and the accompanying contractual arrangements) are *simultaneously* determined on a transaction-by-transaction basis, in accordance with the principle of aligning the transaction's requirements and costs with the available instrument and its capability. Having expounded on the influence of asset specificity on financing choice, in the text that follows, we attempt to interpret the observed patterns of corporate finance and corporate governance in the transport industry in the light of the New Institutional Economics/transaction cost economics. For convenience of exposition, our main focus will be on the example of the ocean transport or shipping industry.

Interaction between fund-raising firm and financier with constraints on financing alternatives and differential costs/benefits

For the firm seeking finance, the first factor that bears on the financing choice is likely to be the opportunity set that is faced, since the financing schemes or instruments available at any time are constrained by the state of economic (including financial market) development (see World Bank, 1997; International Finance Corporation, 1996). It was, for example, impossible for a Chinese shipping company or airline to issue public equity or public debt in the domestic stock market before 1990 since no such market existed![10] But as China tried to reform itself towards a market-like economy (e.g. by establishing two domestic stock exchanges and allowing large firms to be listed overseas), new financing opportunities (most notably, publicly traded securities) were created, and, as a result, the financing of companies also evolved.

In a similar vein, under any given economic condition, the available financing options are largely shaped by factors relating to the type of assets/investments being financed: for example, whether the asset has a liquid second-hand market, the fund-seeker's credit-worthiness and business track record, and the like (see Grammenos and Xilas, 1996).

As documented by Stopford (1997) and Grammenos and Xilas (1996), the shipping industry has consistently relied on commercial bank loans for the majority of their funding needs.[11] The predominant use of private debt versus other financing methods may be explained as follows. Firstly, the equity base of shipping companies has been seriously eroded in the past due to high risk-taking activities and poor product market conditions, which resulted in a lack of internal funds (Grammenos and Marcoulis, 1996).

Secondly, relative to other financing schemes, bank loans provide a quick and convenient source of financing, since the secretive and volatile nature of shipping does not favour public fund-raising due to the secretive nature of the business and the information disclosure requirements (see Stokes, 1996; McConville and Leggate, 1999). The recent experience of Golden Ocean (which got into severe financial distress after a public bond issue in the United States) spoke eloquently of the 'joint ignorance' that exists between shipowners, finance arrangers and public capital providers, and the dire consequences that follow.

Another, yet related, explanation is somewhat less obvious: shipowners actually find it *attractive* to use bank loans. On the one hand, compared with the typically much higher cost of public equity at around 15 to 20 per cent, a bank loan is a much lower cost method of financing, with interest rates at around 1 per cent above LIBOR (the London InterBank Offered Rate) or less for the average borrower. On the other hand, by 'playing' (as in 'asset plays' – the asset being the vessel for which there exists a liquid second-hand market) with money that comes mostly from banks, particularly during times when banks were less prudent, the borrowing shipowner actually has much to gain in a bullish market, but not as much to lose in a bearish market downturn. In effect, the borrower has acquired an option: If the shipping market is good, a handsome return is made (net of debt repayment), since vessel prices are extremely volatile and may rise or drop dramatically within a relatively short span of time.[12] If the freight market turns out bad and vessel prices drop with it, the most that is lost is the shipowner's portion (usually about 30 per cent) of the investment in the ship. In other words, the shipowner may default on debt repayment, leaving the residual value of the ship at the risk and disposal of the bank. The one-company–one-ship ownership structure characteristic of the shipping industry alleviates to some extent the potential impact of failure in one transaction on the borrower's other businesses or reputation. On the other hand, while individual financiers usually remember the lessons they have learned, at any point in time a financial system has a certain proportion of inexperienced agents. This continuing fresh supply of inexperience means that some of the lessons of the past must be learned repeatedly (Neave, 1991: 22). This phenomenon is strongly in evidence in bank shipping finance: banks are known to have unusually short memories, and have been blamed for supplying the industry with easy finance in the past (with some banks sometimes providing up to 100 per cent financing) and for fuelling over-supply, which leads to subsequent market downturns and bad debts. The value/role of the option acquired by shipowners in bank financing warrants a separate study in its own right (but see Uttmark, 2002 for a practitioner's view on this).

From the perspective of the financier, on the other hand, he also has to consider a number of transaction-specific factors. If the transaction is a common financing activity involving predictable risks, then standard forms of rule-based financing and contracts may be used. If, however, the transaction is highly irregular and involves large *ex ante* uncertainty, then a financing/ governance structure must be chosen which is relatively flexible and which allows for sufficient *ex post* adjustments (in contract terms and conditions, etc.). Consider a request from an old customer for (say) a 70 per cent ten-year term loan for a new general purpose dry bulk carrier. In this case, other than predicting the preference of debt over equity, our theory suggests that standard (rule-based) loan arrangements are likely to be used, with the financier usually demanding standard terms and conditions such as charter parties from first-class charterers, the market interest rate, a first-preferred mortgage and collateral securities, and standard-type covenants.

Next, consider a second case, in which a relatively new (but equally creditworthy) client requests (say) a 70 per cent funding for a new tailor-made LPG carrier for which only a 5-year charterparty has been secured. Assume further that the fund-seeker does not insist on any specific type of financing (i.e. he/she is indifferent between debt and equity, whether private or public). Based on the transactional circumstances, our theory will predict that in this case private debt (preferably with a fund supplier with experience in the industry) will be preferred to public debt or public equity. The reason is that not only is the business track record of the new shipowner unknown to the lender, but levels of asset specificity and investment uncertainty are both very high. Equity financiers will have to be given a very high rate of return to entice them into the transaction, and even if this occurs, the restrictions on the company's financial and operating conditions are likely to be demanding. In contrast, debt financing is more likely to work if certain conditions can be worked out between the transacting parties in private. These conditions, other than rate of return and/or fee charges, may include the provision of details about the firm's strategic, financing, and operating aspects, timely and regular provision of the firm's inside information, and more importantly, with flexible contingencies (favourable to the financier) built into the loan contract and allowing for frequent *ex post* adjustments.[13] These are likely to increase the overall transaction costs of the deal, but the funds-hungry firm may have to accept it (happily) because the costs under alternative equity financing schemes are probably even higher, given the asset type and investment risks. The ultimate choice of financing or governance structure will depend on the interaction of various transaction-specific factors, including the perceived trade-off between costs (including disclosure costs) and benefits of the available financing/governance structures, the experience/utility function of the financial intermediary, and the maturity, sophistication and efficiency of the financial system (Neave, 1991).

Country effects and the influence of institutional factors

Not only do financing/governance structures differ systematically across industries (as characterized by the similarity of investment assets and opportunities) and over time (as a result of the evolution of the wider socioeconomic environment), but there also appears to be country-specific characteristics in the financing of the same industry at any given time, a result that is attributable to different institutional features in different countries. Booth et al. (2001) analyse capital structure choices of firms in ten developing countries, and provide evidence that these decisions are affected by the same variables as in developed countries. However, there are persistent differences across countries, indicating that although some of the insights from modern finance theory are portable across countries, much remains to be done to understand the impact of different institutional features on capital structure choices.

In the air transport sector, local or central government funding has essentially financed airports in most parts of the world. Even though this system will no longer be able to cope with the large investments required to upgrade the world's airports, various institutional barriers exist in many countries that deny airports access to other funding sources. In the United States, for instance, public airports have access to tax-free debt through revenue bonds, making debt relatively cheap, even though it may be limited in available volume (Doganis, 1992). In some other countries, the institutional structure of the airports makes it difficult or even impossible for the use of private finance unless institutional changes in the legal framework of airport ownership or tax regulation are implemented first. Such changes are always difficult and slow to bring about (Doganis, 1992). Until such barriers are cleared, the financing pattern in the sector is shaped mainly by institutional, political and economic factors. The immediate effect of institutional changes on financing behaviour is evidenced in the public listing of two of China's major airlines, China Eastern Airlines and China Southern Airlines, in both local and overseas stock exchanges after the regulatory barriers were removed in the early 1990s. The changes in these airlines' financing choices is a direct result of the country's economic reforms, under which some formerly state-owned enterprises are allowed to seek public listing.[14]

Applications of the TCE-based approach to other forms of economic organization

Williamson (1988) shows that the general TCE asset-based approach may be applied to examine the economic rationale for a variety of other forms of economic phenomena, including leasing and leveraged buyouts. The common theme is that they can all be traced back to the same TCE imperative, which is to align transactions (which differ in their attributes) with governance structures (the costs and competencies of which differ) in a discriminating (mainly, transaction-cost-economizing) way.

The TCE justification for leasing as a governance structure is as follows. Assume that standby access to an asset on wheels is required and that market procurement of the services supplied by this asset is believed to be defective.[15] If the assets in question are durable, general-purpose assets and they are resistant to user abuse (and/or the costs of inspection and attributing abuse are low), then leasing may evolve as a natural response.

This may be explained as follows. Firstly, let k be an index of asset specificity and let D(k) and E(k) represent the cost of debt and equity capital, respectively. It is easy to show that, at very low asset specificity, the cost of debt, D(0), is necessarily lower than the cost of equity, E(0), since the former is a relatively simple, rule-governed relation whereas the latter is a much more complex governance relation involving higher set-up costs and intensive monitoring/politicking. As asset specificity deepens, the costs of both debt and equity finance increase, but debt financing rises more rapidly (i.e. $D' > E'$). This is because a rule-governed regime will sometimes force liquidation or otherwise cause the firm to compromise value-enhancing decisions that a more adaptable regime, of which equity governance is one, could implement. Let $\bar{k}$ be the value of k for which E(k) = D(k). Then the optimal choice of all-or-none finance is to use debt for all projects for which $k < \bar{k}$ and equity finance for all $k > \bar{k}$.

Next, based on the impossibility of 'selective intervention' (mainly because of managerial discretion/agency costs), Williamson (1988) shows that an intermediate form of governance (called 'dequity', which combines the strengths of discretion-based equity and those of rule-based debt) cannot dominate both debt and equity over the full range of k. This intermediate form of governance (δ) has the following properties: $D(0) < \delta(0) > E(0)$; and $D' > \delta' > E' > 0$.

General-purpose assets (e.g. ships, airplanes, railroad cars, trucks) satisfy the condition k = 0. Moreover, since measurement problems are assumed to be negligible, there is no need to combine owner and user for user-cost reasons. Since an outside owner that is specialized in this type of equipment (or that has a competitive advantage in handling this type of equipment) can repossess and productively redeploy these assets more effectively than could a more specialized debt-holder, leasing is arguably the form of finance that has the least overall costs for this type of asset.

It is interesting to compare the above analysis with those in section 4.1, in particular with Macquarie Corporate Finance Limited's (2000) explanation of the popular use of leasing for standardized transport assets, and with the empirical results of Palay (1984). It is also useful to contrast TCE's approach to leasing with that in corporate finance. Brealey and Myers (2000) summarize 'sensible' reasons for leasing as follows:

- Short-term leases are convenient (e.g. renting a car for a week on holiday; obtaining an operating lease on equipment for a year or two).
- Cancellation options are valuable.
- Maintenance is provided (in the case of a full-service lease).
- Standardization leads to low administration and transaction costs.
- Tax benefits (e.g. depreciation allowances for lessor; tax deductible and/or low lease payments).

It can be seen that both corporate finance and TCE provide similar (and complementary) justifications for the use of a lease for standardized equipment. While corporate finance places more emphasis on 'valuation effects' (e.g. tax benefits and the value of flexibility), TCE enjoys the advantage of being a consistent, unified, transaction-cost-minimizing approach. Thus, combining insights from both modern finance and transaction cost economics/institutional economics add to our understanding of economic organization.

It is possible to extend the above analysis not only to project finance (e.g. toll roads) and bareboat chartering (a practice similar to a financial lease), but also to mergers and acquisitions (M&As) and alliances. While

M&As (as well as leveraged buyouts) have been treated by Williamson and others using TCE reasoning, alliances have not yet been subjected to similar analyses, at least in a transport industry context. Panayides and Gong (2002) review consolidation, mergers and acquisitions in the shipping industry. One of the questions they raise is: why do some shipping companies prefer M&As to alliances? It is our conjecture that this may be because a full M&A is regarded as a governance structure through which the intended benefits (e.g. wider market coverage, operating and financial synergies) of the strategic move can be realized at lower overall costs (e.g. the cost of coordination, which is higher in an alliance). Future studies may be able to shed more light on this.

In summary, corporate finance and governance is obviously influenced by the wider economic and institutional environment, the costs and competences of different financing/governance structures, as well other transactional circumstances such as asset specificity and the informational condition. While the implications of the theory proposed herein need to be systematically tested, it is comforting to see that the predictions of our theory are broadly consistent with observed practice in the industry, and this is not just because we have developed the theory on the basis of the observations themselves. In fact, it is in the evaluation of the usefulness of the theory that lies the key difference between the transactions cost economics approach and the neoclassical finance approach to economic organization. A descriptive theory of the type advanced here, owing to its focus on an individual transaction as the basic unit of analysis, may be said to suffer from being too much tied to a transaction or a class of transactional circumstances, leading to a lack of generalizability. But this latter criticism is clearly unjustified, since, as Williamson (1988) and others show, the transaction economics approach actually applies to a wide class of economic organization problems not limited to corporate finance. In contrast, sweeping statements about financing behaviour at the aggregate level that are results of studies which abstract from real-life complexities are probably less likely to offer similar insights (obvious as they sometimes are) that are consistent with real-life practices. The usual caveat is thus in order: the trade-off between eclecticism and wide generalizability may depend on the intended use to which the theory is put.

We believe that more systematic evidence to be provided in future research will be able to extend greater support for our proposition, which may be summarized as follows: The corporate finance and corporate governance structure in the transport industry is jointly determined in a total-cost-minimizing way by transactional characteristics such as asset specificity and product market characteristics, and by the interaction between the entity seeking finance and the financier, under the constraint of available financing alternatives (themselves a function of the wider socioeconomic system).

4.6   Conclusion and directions for future research

Modern finance theories, which are non-institutional in nature, suggest that capital structure is irrelevant or that there exists a pecking order according to which corporate financing choices are made. While the pecking order theory is able to generate predictions that are broadly consistent with some observed corporate financing practices, it appears incapable of providing non-ambiguous justification for many categories of financing behaviour observed in general industry, and in the world transport industry in particular. There is strong evidence to suggest that in real life, there exist wider economic, institutional and industry-specific factors that exert an important influence on the way in which an industry is financed. There are also clear cross-sectional patterns in the financing methods in different sub-sectors of the transport industry, and such patterns are related to asset specificity, industry characteristics, and other institutional as well as non-institutional factors.

In this chapter we have proposed an eclectic approach to corporate finance and corporate governance in the world transport industry by combining insights from neoclassical financial economics, the New Institutional Economics (of which transaction cost economics (TCE) is a fast-growing branch), and development economics.

Although such a micro-analytic approach may be criticized as being too narrow or industry-specific, the synergy thus created generates an analysis that is more consistent with real-life observations than can be achieved by relying on any single paradigm. Given the dearth of studies in the transport industry that systematically explore the economic and institutional rationale behind observed financing practices, it behoves future researchers to conduct a series of coordinated research studies along the lines of the descriptive theory we have proposed herein. Other than providing empirical evidence on the past and current use of different financing methods in the industry, future research on transport financing may also examine the comparative cost-effectiveness and capability of different types of financing and governance structures (e.g. public debt or equity versus private debt or equity). Regarding these governance structures as more than financial (e.g. risk management) instruments may lead us to examine shipping pools, alliances, and hybrid financing schemes (e.g. leasing, bareboat chartering etc.) through a unified, TCE-based approach of economic organization. Outcomes of such analyses may be of interest not only to academic researchers and industry practitioners; it may also inform governments in the formulation of economically efficient and effective policies that are geared towards building a strong transport infrastructure. From another perspective, the results from such focused studies may also contribute to the refining of existing finance and economic theories.

Notes

1 This discussion about TCE draws heavily from Williamson (1988).
2 Myers (1984) noted in footnote 3 that Modigliani's and Miller's (1966) use of the electric utilities industry enables them to side-step the issue of controlling for other variables (such as a tax-saving incentive for using debt, since the utilities can pass these costs to customers) in their study of capital structure. This is yet another example of the advantages of adopting an industry-focused approach (needless to say, such advantages depend on the purpose of the investigation), although the M&M study has sometimes been criticized for its focus on a single industry.
3 This proposition is in fact the reason why investment and financing decisions can be completely separated (Brealey and Myers 1991: 397). It is now generally recognised that financing decisions are in fact related to the investment decisions and in turn to the product market environment (see Long and Malitz, 1985; Spence, 1985; Maksimovic, 1990).
4 When there is tax, M&M's proposition I as 'corrected' in their 1963 paper suggests that firms should borrow as much as they can, without, of course, over-looking the potential cost of financial distress. Miller (1976) provides a treatment of capital structure under differential tax rates. See Brealey and Myers (2000) for a summary of these two papers.
5 But even this may be consistent with the static trade-off hypothesis: profitable firms might find that the financial benefits that can be derived from the use of debt are outweighed by the attendant costs (of information disclosure, restrictive covenants, etc.). In addition, no theory can be expected to be entirely consistent with all observations. The choice of alternative theories may have to be arbitrary to some extent, and depend upon the purpose of the investigation. We prefer to trade off 'simplicity' for 'fruitfulness' in our choice among alternative theories, since taking account of a wider spectrum of economic (including industry) factors results in more precise predictions. The disadvantage is that the conclusions may not be generalized to other industries or circumstances.
6 Evidence of companies aligning financing requirements with the costs of different financing instruments on a transaction-by-transaction basis has important implications for capital budgeting. For, if the cost of each project is different, then across-the-board use of the company's weighted average cost of capital (WACC) would be unwarranted.
7 Gong, Firth and Cullinane (2001) report that about 800 transportation companies have had their shares listed on major stock exchanges *worldwide* as of December 2000. Although the size of each of public equity and public debt is unknown, the former is believed to be far greater than the latter.
8 But see Subrahmanyam and Titman (1999) for a treatment of the linkages between stock price efficiency, the choice between private and public financing, and the development of capital markets. Among others, they find that the advantage of public financing is high if costly information is diverse and cheap to acquire. The effects of an information disclosure cost differential between different forms of financing on the choice of financing sources are also discussed in Yosha (1995).
9 See Coase (2000) for a definition of the New Institutional Economics.
10 In the words of a pundit, a choice is a choice only if there is a choice.

11   The Chinese shipping companies are probably one of the rare exceptions. Until the early 1990s, Chinese shipping companies had been funded primarily (if not solely) out of the state budget. A similar case seemed to have applied in the Former Soviet Union (see Chrzanowski, 1975).
12   Many shipowners have admitted on various occasions that they make much more money from 'buy-low–sell-high' investment strategies ('asset plays') than from carrying cargoes.
13   Leland and Pyle (1977) show that financial intermediation can be viewed as a natural response to asymmetric information, since these intermediaries enjoy economies of scale in acquiring and processing private information for certain classes of assets for which information is not publicly available or not credible, a situation that is quite true of the infamously secretive shipping industry.
14   For more international evidence, see Zhang (1998), Ohta (1999), and Hayashi et al. (1998).
15   We believe that information asymmetries and moral hazard are particularly important reasons for making the market mode inefficient. Specifically, insofar as the shipping industry is concerned, the secretive and risk-laden nature of the business, combined with the lack of credible public (financial) information about the industry firms, makes it very expensive to use public funds.

References

Allen, F. and Winton, A. (1995) Corporate Financial Structure, Incentives and Optimal Contracting, in R.A. Jarrow, V. Maksimovic and W.T. Ziemba (eds), *Handbooks in Operations Research and Management Science*, vol. 9 Finance, Amsterdam: North-Holland.
Baskin, J. (1989) An Empirical Investigation of the Pecking Order Hypothesis, *Financial Management*, 18, 26–35.
Booth, L., Aivazian, V., Demirguc-Kunt, A., and Maksimovic, V. (2001) Capital Structure in Developing Countries, *Journal of Finance*, 56 (1), 87–130.
Brealey, R.A. and Myers, S.C. (1991) *Principles of Corporate Finance*, 4th edition, New York: McGraw-Hill Inc.
Brealey, R.A. and Myers, S.C. (2000) *Principles of Corporate Finance*, 6th edition, New York: McGraw-Hill Inc.
Chrzanowski, I.H. (1975) *Concentration and Centralization of Capital in Shipping*, Lexington, MA: Lexington Books.
Coase, R.H. (2000) The New Institutional Economics, in C. Menard (ed.), *Institutions, Contracts and Organizations: Perspectives from New Institutional Economics.* Cheltenham: Edward Elgar.
Cullinane, K.P.B. and Gong, X. (2002) The Mispricing of Transportation Initial Public Offerings in the Chinese Mainland and Hong Kong, *Maritime Policy & Management*, 29(2), 107–18.
Doganis, R. (1992) *The Airport Business*, London: Routledge.
Donaldson, G. (1961) *Corporate Debt Capacity: A Study of Corporate Debt Policy and the Determination of Corporate Debt Capacity*, Division of Research, Graduate School of Business Administration, Harvard University, Cambridge, MA.
Drewry (1999) *Shipping Finance: A High Risk–Low Return Business*, London: Drewry Consultants.
Elton, E.J. and Gruber, M.J. (1970) Homogeneous Groups and the Testing of Economic Hypotheses, *Journal of Financial and Quantitative Analysis*, 4, 581–602.

Fama, E.F. and Miller, M.H. (1972) *The Theory of Finance*, Hinsdale, IL: Dryden Press.

Friedman, B.M. (1985) Corporate Capital Structures in the United States: An Introduction and Overview, in B.M. Friedman (ed.), *Corporate Capital Structures in the United States*, Chicago and London: University of Chicago Press.

Friedman, M. (1953) *Essays in Positive Economics*, Chicago and London: University of Chicago Press.

Gong, X., Firth, M., and Cullinane, K.P.B. (2001) An Empirical Examination of the Price Performance of Transportation Common Stocks, *Proceedings of the 9th World Conference on Transport Research*, Seoul, Korea.

Grammenos, C.T. and Marcoulis, S.N. (1996) Shipping Initial Public Offerings: A Cross-Country Analysis, in M. Levis (ed.), *Empirical Issues in Raising Equity Capital*. Amsterdam: North-Holland.

Grammenos, C.T. and Xilas, E.M. (1996) *Shipping Investment & Finance*, Course manual, Department of Shipping, Trade and Finance, City University Business School, London.

Hayashi, Y., Yang, Z.Z. and Osman, O. (1998) The Effects of Economic Restructuring on China's System for Financing Transport Infrastructure, *Transportation Research A-Policy and Practice*, 32A(3), 183–195.

International Finance Corporation (1996) *Financing Private Infrastructure*, Washington, DC: The World Bank.

Jensen, M. and Meckling, W. (1976) Theory of the Firm: Managerial Behaviour, Agency Costs, and Capital Structure, *Journal of Financial Economics*, 3, 305–60.

Joskow, P. (1988) Asset Specificity and the Structure of Vertical Relationships: Empirical Evidence, *Journal of Law, Economics, and Organisation*, 4, 95–177.

Klein, B., Crawford, R.G. and Alchian, A.A. (1978) Vertical Integration, Appropriable Rents, and the Competitive Contracting Process, *Journal of Law and Economics*, 21(2), 297–326.

Langner, S.J. (1995) Contractual Aspects of Transacting in Slots in the United States, *Journal of Air Transport Management*, 2, (3–4) 151–61.

Leland, H.E. and Pyle, D.H. (1977) Information Asymmetries, Financial Structure, and Financial Intermediation, *Journal of Finance*, 32(2), 371–87.

Lee, T.W. (1996) *Shipping Developments in Far East Asia: The Korean Experience*, Aldershot: Avebury.

Long, M.S. and Malitz, I.B. (1985) Investment Patterns and Financial Leverage, in B.M. Friedman (ed.), *Corporate Capital Structures in the United States*, Chicago and London: University of Chicago Press.

Macquarie Corporate Finance Limited (2000) *The Guide to Financing Transport Projects*, London: Euromoney Books.

Maksimovic, V. (1990) Product Market Imperfections and Loan Commitments, *Journal of Finance*, 45(5), 1641–53.

Maksimovic, V. (1995) Financial Structure and Product Market Competition, in R.A., Jarrow, V. Maksimovic, and W.T. Ziemba (eds), *Handbooks in Operations Research and Management Science*, vol. 9 Finance, Amsterdam: North-Holland.

McConville, J. and Leggate, H.K. (1999) Bond Finance for the Maritime Industry, paper presented at the International Association of Maritime Economists Annual Conference, Halifax, Canada, 13–14 September .

Miller, M.H. (1976) Debt and Taxes, *Journal of Finance*, 32, 261–76.

Miller, M.H. (1988) The Modigliani–Miller Propositions after Thirty Years, *Journal of Economic Perspectives*, 2(4), 99–120.

Miller, M.H. and Modigliani, F. (1966) Some Estimates of the Cost of Capital to the Electric Utility Industry, 1954–57, *American Economic* Review, 56, 333–91.

Modigliani, F. (1982) Debt Dividend Policy, Taxes, Inflation and Market Valuation, *The Journal of Finance*, 37(2), 255–73.
Modigliani, F. (1988) MM – Past, Present, and Future, *Journal of Economic Perspectives*, 2(4), 149–58.
Modigliani, F. and Miller, M.H. (1958) The Cost of Capital, Corporation Finance and the Theory of Investment, *The American Economic Review*, 48, 261–97.
Modigliani, F. and Miller, M.H. (1963) Corporate Income Taxes and the Cost of Capital: a Correction, *American Economic Review*, 53, 433–43.
Monteverde, K. and Teece, D.J. (1982) Appropriable Rents and Quasi-Vertical Integration, *Journal of Law and Economics*, 25, 321–8.
Myers, S.C. (1984) The Capital Structure Puzzle, *Journal of Finance*, 39, 575–92.
Myers, S.C. and Majluf, N.S. (1984) Corporate Financing and Investment Decisions When Firms Have Information Investors do not Have, *Journal of Financial Economics*, 13, 187–222.
Neave, E.H. (1991) *The Economic Organisation of a Financial System*, London: Routledge.
Ohta, K. (1999) International Airports: Financing Methods in Japan, *Journal of Air Transport Management*, 5(4), 223–34.
Palay, T.R. (1984) Comparative Institutional Economics: The Governance of Rail Freight Contracting, *Journal of Legal Studies*, 13, 265–87.
Panayides, P.M and Gong, X. (2002) Consolidation, Mergers and Acquisitions in the Shipping Industry, in C.T. Grammenos (ed.), *The Handbook of Maritime Economics and Business*, London: Lloyd's of London Press.
Rybczynski, T.D. (1984) Industrial Financial System in Europe, U.S. and Japan, *Journal of Economic Behaviour and Organisation*, 5, 275–86.
Shelanski, H.A. and Klein, P.G.. (1995) Empirical Research in Transaction Cost Economics: a Review and Assessment, *Journal of Law, Economics, and Organization*, 11, 335–61.
Simon, H.A. (1957) *Models of Man*, New York: John Wiley & Sons.
Spence, A.M. (1985) Capital structure and the corporation's product market environment, in B.M. Friedman (ed.), *Corporate Capital Structures in the United States*, Chicago and London: University of Chicago Press.
Stokes, P. (1992) Ship Finance: *Credit Expansion and the Boom–bust Cycle*, London: Lloyd's of London Press.
Stokes, P. (1996) Problems Faced by the Shipping Industry in Raising Capital in the Securities Markets, *Maritime Policy and Management*, 23(4), 397–405.
Stopford, M. (1997) *Maritime Economics*, London: Routledge.
Subrahmanyam, A. and Titman, S. (1999) The Going–Public Decision and the Development of Financial Markets, *Journal of Finance*, 54(3), 1045–82.
Swoboda, P. and Zechner, J. (1995) Financial Structure and the Tax System, in R.A., Jarrow, V. Maksimovic and W.T. Ziemba (eds), *Handbooks in Operations Research and Management Science*, vol. 9 Finance, Amsterdam: North-Holland.
Taggart, R.A. (1985) Secular Patterns in the Financing of U.S. Corporations, in B.M.Friedman (ed.), *Corporate Capital Structures in the United States*, Chicago and London: University of Chicago Press.
Uttmark, G.. (2002) Victory!, *Marine Money*, 18(4), 23–5.
Welch, I. (1995) A Primer on Capital Structure, *Finanzmarkt und Portfolio Management*, 2, 232–49.
Williamson, O.E. (1979) Transaction Cost Economics: The Governance of Contractual Relations, *Journal of Law and Economics*, 22, 223–61.

Williamson, O.E. (1988) Corporate Finance and Corporate Governance, *Journal of Finance,* 43(1), 567–92.

Williamson, O.E. (1993) *The Economic Analysis of Institutions and Organisations: In General and with Respect to Country Studies,* Paris: OECD.

Williamson, O.E. (1996) *The Mechanisms of Governance,* New York: Oxford University Press.

Williamson, O.E. and Masten, S.E. (1999) *The Economics of Transaction Costs,* Cheltenham: Edward Elgar.

World Bank (1997) *Mobilizing Domestic Capital Markets for Infrastructure Financing: International Experience and Lessons for China,* World Bank Discussion Paper No. 377, Washington, DC: World Bank.

Yosha, O. (1995) Information Disclosure Costs and the Choice of Financing Source, *Journal of Financial Intermediation,* 4, 3–20.

Zhang, A.M. (1998) Industrial Reform And air Transport Development in China, *Journal of Air Transport Management,* 4(3), 155–64.

5

A Fuzzy Set Theory Approach to Flagging Out: Towards a New Chinese Shipping Policy[1]

Hercules E. Haralambides and Jiaqi Yang[2]

5.1 Introduction

Changing a vessel's registry from an 'expensive' to a 'cheap flag' has long been a worldwide phenomenon. This 'internationalization' strategy has been adopted primarily by traditional (*developed*) maritime countries (TMCs) for reasons of reducing operating costs and fiscal pressures, as well as for enhancing overall operational flexibility.

Over the years, since the inception of the institution of open registries in the 1950s, a substantial number of studies of this phenomenon have been undertaken, but their overriding focus on TMC policies has not always provided an adequate example to the different 'realities' of the new developing countries (NDCs), such as China, who are also increasingly adopting this policy.

The flagging out of Chinese-controlled vessels started to occur, for a number of mainly political reasons, at the same time as in other nations (1950s), but it began to accelerate following China's opening to the outside world in the 1980s, and particularly since the reform of the country's tax regime in the 1990s. As in many other countries, this evolution is becoming increasingly serious, in terms of its negative impacts on the country's economic development, particularly as flagging out continues unchecked. Given the variety and eligibility of positive measures, taken predominantly in the European Union, in order to reverse similar trends, China could by no means be accused of being 'protectionist' were she to adopt more preferential shipping policies in order to reflag vessels to her national registry.

This chapter consists of five sections. The first section discusses the background and objectives of the study and reviews the pertinent literature. The second section reviews international and Chinese experiences and developments in flagging out, simultaneously considering the impacts of flagging out on international shipping, national economies and society. China's motivations for flagging out are also addressed here. The objective of the third section is to identify the determinants of 'adjustment' in shipping policy. Here, our chapter introduces fuzzy set theory and related models – as well as an international questionnaire survey – to assess the economic effects of flagging out through context-dependent economic and societal indicators (factors). The following section is based on the determinant analysis of fuzzy models and mainly focuses on the 'policy competition' and 'government intervention' policies to counteract flagging out. In this context, a comparative shipping policy analysis is made between China and TMCs, in order to evaluate the degree of openness of China's shipping policy and explore possible policy alternatives. Some concluding comments follow in the last section.

5.2   The evolution and impact of flagging out

The use of the flag of convenience (FOC) institution dates back to 1922 when the United States took the initiative to permit the registration of American ships in Panama. Presently, as a result of lower crewing costs, tax exemptions and minimal bureaucracy, the greatest proportion of the world merchant fleet is registered under FOCs (Figure 5.1). At the beginning of 2000, 12,996 merchant ships with 442.1 million dwt, or 61.8 per cent of world tonnage, were not registered in the country of domicile of the owner (ISL, 2000).

The share of tonnage beneficially owned by developing countries has also increased continuously since 1980, reaching one-third of FOC tonnage in 1998. According to UNCTAD, the major developing maritime countries and territories – including China, Hong Kong (China), Republic of Korea, Saudi Arabia, UAE etc. – had more than half of their tonnage registered under foreign flags.

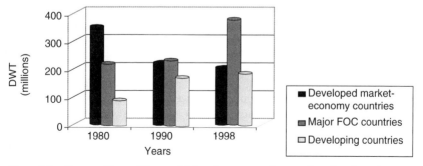

Figure 5.1   Composition of the world fleet by groups of countries of registration
Source: UNCTAD (1999).

5.3   Flagging out in China

Evolution and current status

Chinese shipping has witnessed strong growth in recent decades. This was the result of rapid industrialization, economic growth and trade. By the end of 1998, more than 300 companies were registered with the Ministry of Communications as offering international shipping services (MoC, 1999). The number of vessels registered in China was 1,880, corresponding to 16 million dwt. This capacity meant China ranked ninth in the world – or fifth if FOC registered tonnage is included. In the same year, the latter tonnage comprised 550 ships of 20.15 million dwt. Clearly, in terms of dwt, more than half of all Chinese-owned or -controlled tonnage was registered under FOCs and this trend is continuing. In 1991, the share of FOC tonnage was only 23 per cent.

Chinese flagging out started in the 1950s. As a result of western embargo policies, Chinese trade was carried by joint venture companies set up with socialist partners such as Poland, Czechoslovakia and Albania. By the 1960s, China had built up its own fleet, under its national flag, although a small number of vessels were still flagged out in order to seize trading opportunities with countries with which China had not, as yet, established diplomatic relations.

Flagging out took massive dimensions and momentum with the opening up of the country in the 1980s, especially since China's tax reforms in 1994. As in all other cases, profitability and operational considerations were the prime movers in the flagging-out process. A number of such considerations could be mentioned, including:

- High tariffs and value added taxes (27 per cent) on imported ships and
- Industrial carriage.

Shippers registered their own tonnage abroad to avoid stringent company laws, shipbuilding and trading regulations and other limitations imposed on national flag ships. For instance, in 1998, the Beijing-based *Capital Iron and Steel Group* owned 1.1 million FOC tonnage with future plans to develop an additional modern bulk cargo fleet, exceeding 4 million dwt, in cooperation with a British shipping enterprise. This fleet was also intended to be registered abroad. Several other Chinese manufacturing enterprises are expected to build up their own ocean-going FOC fleets in the coming years.

In terms of composition (Table 5.1), the largest part of the Chinese fleet consists of bulk carriers (57.2 per cent), followed by tankers (13.1 per cent), container ships (11.8 per cent), general cargo ships (7.4 per cent) and multipurpose vessels (6 per cent).

Bulk carriers, tankers and container ships are also the sectors where flagging out is most pronounced (63.4 per cent, 61.1 per cent, and 53.7 per cent respectively, all higher than world averages; see Figure 5.2). The same world pattern is observed with regard to flag preference, with Panama, Liberia, Malta and Cyprus leading the list (Table 5.2).

Despite conventional thinking, ships registered under national flags are not necessarily younger than FOC vessels, and this is particularly true for such countries as Japan, the United States, China and the Republic of Korea. At the beginning of 2000, ships under national registers had an average age of 17.6 years compared to 15 years of FOC vessels. Only ships registered under the flags of Germany and Switzerland were, primarily as a result of taxation rules, on average much younger than their flagged-out counterparts (ISL, 2000).

The average age of the Chinese ocean-going fleet was 18.5 years, at the end of 1998. Amongst them, passenger/container ships, crude oil tankers,

Table 5.1   Chinese fleet composition as at the end of 1998

Ship type	DWT	Average age (years)	% of Chinese flag
Bulk carriers	20,876,521	13.98	36.6
Oil tankers	5,437,851	12.10	38.9
Container ships	4,305,305	12.34	46.3
Multipurpose	2,200,051	13.80	63.8
General cargo	2,686,928	21.90	93.4
LNGs	82,123	20.10	100
Passenger ships	64,317	16.80	78.1

Source: MoC (1999).

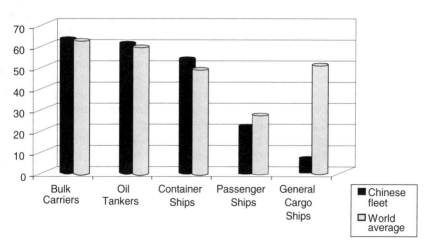

Figure 5.2   Proportion of Chinese and world FOC vessels by ship type
Source: China Shipping Development Annual Report (1999), MoC and ISL Market Analysis 2000.

Table 5.2   Flag composition of Chinese FOC fleet (as of 1 January 1996 and 2000)

Open registry flags	1996	2000
Panama	37.4	50.0
Liberia	37.2	20.6
Malta	3.5	1.9
Cyprus	1.7	1.2
Marshall Islands	0.9	—
Saint Vincent	7.2	7.5
Cayman Islands	1.3	—
Others	10.8	18.8
Total	100	100

Source: ISL Market Analysis 2000.

chemical vessels, reefers and container ships (excluding cellular) were the youngest with an average age between 10 and 12 years. Passenger/cargo vessels, passenger/vehicle Ro-Ro, break bulk and LPG carriers were among the oldest with an average age of 20 years. By contrast, the average age of FOC vessels under Chinese control was 12 years. Among them, passenger/ container vessels were on average only 2.9 years old, and the average age of passenger/vehicle Ro-Ro vessels, chemical vessels, oil tankers, multipurpose vessels and full container vessels was from 9.2 to 13.6 years.

The impact of Chinese flagging out

The widespread flagging-out phenomenon, affecting not only the merchant fleet of the vast majority of the traditional maritime countries but also the new developing maritime nations, has attracted a great deal of attention for a variety of reasons (Bergantino and O'Sullivan, 1999). One of the main concerns has been the fact that open registry fleets have expanded at a faster rate than that recorded by any other fleet in the world and, as a result, this expansion has limited the growth of national fleets, with all the related consequences for national defence, fiscal revenue and the disappearance of national seafarers. In addition, the occurrence in recent years of several alarming incidents involving environmental disasters has increased public awareness of this problem.

In China, flagging out has not only exerted a great influence on the 'cohesiveness' of the national ocean-going fleet, but it has also caused a series of problems to national macroeconomic control, tax revenues and the employment of seafarers.

As mentioned above, imported vessels are liable to import tariffs and value-added tax, totalling 27.53 per cent of the price of the vessel. Such taxes have forced a number of shipowners to build or purchase vessels abroad and register them under foreign flags. This has deprived the country not only of tariff and tax revenues, but also of registration and inspection fees. As a result of ineffective control over the flagged-out ships, corporate taxes may also be lost to China. It is worth noting here that approximately US$5 billion of shipping revenues per year are deposited and spent abroad, with all evident consequences on national foreign exchange reserves and the ability of the country to repay in foreign exchange.

5.4 Comprehensive assessment of the flag choice decision: a fuzzy evaluation model

General analysis

The decline in nationally owned and registered fleets in recent years has led many European countries to establish 'captive' international or second registers. Examples are UK's Isle of Man, France's Kerguelen Islands, Norway's NIS (Norwegian International Register), GIS in Germany, and DIS in Denmark.

International registers such as these are often seen as placed in that 'grey area' between traditional registers, where legal, economic and administrative links are direct and tight, and open registers where they are very weak (Veenstra and Bergantino, 2000). Seen in this way, they can serve the dual purpose of retaining some control over the national shipping industry and, at the same time, satisfying the need for a more competitive environment on the part of shipowners.

For both shipowners and managers, flag choice is a high-level decision involving a number of external experts such as lawyers, bankers, charterers, manning agencies and classification societies. Normally, shipowners and managers use a mixture of flags for their fleet although they generally have a 'preferred flag'. To a certain extent, shipowners and managers rely on experience and subjective views when making flagging decisions. Quite often, such decisions are frequently readjusted on the basis of such fuzzy perceptions as 'availability of skilled labour', 'maintenance and safety requirements', 'public relation considerations', 'likes and dislikes' and so on. Clearly, considerations such as these augment substantially the complexity of the choice of flag decision.

To a certain extent, the shipping industry, seen in its entirety, constitutes a highly complex system characterized by uncertainty, both in structure and in measurement accuracy. In systems like these, piece-meal deterministic approaches to explaining individual behaviour (such as the flagging decision) are often inadequate. Stochastic (probabilistic) modelling of the overall system, despite its complexity, can achieve much more but, even here, there is a limit to the extent that fuzzy perceptions like the above could be taken into account.

The methodology suggested below addresses such concerns through the use of fuzzy set theory. This approach is able to link human perceptions (say, on the issue of flag selection), expressed in verbal propositions, with numerical measurement (indicators). The merit of the method is that it is completely independent of the model structure, which, in most cases, is only known with little degree of certainty.

Fuzzy set theory was developed by Zadeh in 1965 to deal with imprecise and uncertain problems, which have no well-defined, unambiguous meaning (Cornelissen et al., 2000). The theory has been applied to complex (economic, societal, etc.) decision problems that can be controlled by humans but are hard to define with any precision (Mansur, 1995).

Ordinary (classical) sets are based on binary logic. For instance, membership of the group of 'tall' people (the universal set) can only take two values: 1 for tall and 0 for 'not tall'. Accordingly, a *membership function* simply assigns values of 0 or 1 to each individual element in the universal set. Simply, the membership function just distinguishes between non-members and members of the *crisp* set by a hard threshold. Middle values or partial memberships are not included in the crisp set (Mansur, 1995). However, a hard (crisp) threshold is often unrealistic in practice, because two nearly indistinguishable measurements on either side of the hard threshold will be placed in complementary subsets (Cornelissen et al., 2000).

Contrary to classical set theory, fuzzy set theory is based on multi-valued logic. Let the universe of discourse U have a fuzzy set A described by a membership function μ_A that takes values in the interval $(0, 1)$, μ_A: U $(0, 1)$. Thus, A can be represented by: $A=\mu_A (x)/x$, where $x \in$ U. The membership function μ_A defines a partial membership in a set. This means that μ_A assigns to each x a value from 0 through 1, indicating the degree to which x belongs to A. Transition between membership and non-membership, therefore, is gradual rather than abrupt.

The following part shows how to develop comprehensive fuzzy evaluation models in order to assess the effects of flag choice, based on context-dependent economic, political and societal indicators or factors. Evidently, the methodology is able to define the degree to which such indicators contribute to the choice of flag decision.

Flag choice: qualitative analysis and fuzzy assessment

The comprehensive fuzzy evaluation (CFE) model proposed here is based on fuzzy set theory as developed by Zadeh and on the analytic hierarchical process developed by Saaty (1980). Zadeh defined fuzzy logic as 'the logic underlying models of reasoning which are approximate rather than exact' (Mansur, 1995). Saaty advocated the use of a deductive and systems approach in the analysis of complex decision problems. Along these lines, the scheme of a comprehensive fuzzy evaluation model, to assess the effects of flag choice, is depicted in Figure 5.3.

Six steps are involved: Step 1 defines model input; the set of judgement factors U_i. These fall into three categories: economic, societal and political.

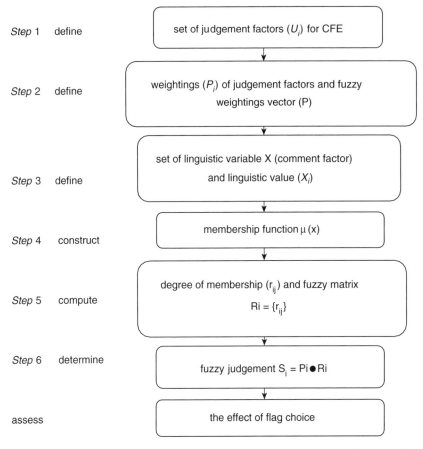

Step 1   define — set of judgement factors (U_i) for CFE

Step 2   define — weightings (P_i) of judgement factors and fuzzy weightings vector (P)

Step 3   define — set of linguistic variable X (comment factor) and linguistic value (X_i)

Step 4   construct — membership function μ (x)

Step 5   compute — degree of membership (r_{ij}) and fuzzy matrix $Ri = \{r_{ij}\}$

Step 6   determine — fuzzy judgement $S_i = Pi \bullet Ri$

assess — the effect of flag choice

Figure 5.3   Comprehensive fuzzy evaluation model to assess the effects of flag choice

Economic factors consist of operating/manning costs; capital costs; maintenance and safety compliance costs; tax-related expenses; the age and size of vessels; fiscal reasons, and so on. It has been shown that, in the EU, manning or crew costs and corporate taxes are the most flag-sensitive cost categories, while capital costs seem to influence the decision to change flag only marginally. For a Chinese shipowner, however, capital costs are perhaps the single most crucial factor in the flagging decision: high import tariffs and value added taxes, coupled with limited credit lines and unfavourable loan conditions for domestically built vessels, can easily explain this.

Societal and political factors in the flagging decision are, however, much more intricate and difficult to assess than the relatively simple capital cost calculations above. These include such attributes as safety standards and requirements; protection of the environment; national defence; labour quality and availability; degree of control; level of bureaucracy; trade union considerations; the general economic and social situation of the home country; the country's industrial structure; public relations; trade restrictions (e.g. embargoes); shippers' preferences, and so on.

As already mentioned, political considerations are also still present, albeit to a lesser extent nowadays. These can be seen, for instance, in the direct seaborne trade between the Chinese mainland and Taiwan: according to existing regulations drawn up by both sides of the Taiwan Straits, only Chinese (mainland or Taiwan) vessels flying FOC flags are allowed to trade on the direct route across the Taiwan Straits.

To assess how the above factors can influence flag choice through the CFE model developed here (i.e. the choice between 'national registry', 'second registry' and 'open registry'), a survey was conducted among more than 100 shipowners in major maritime countries. The survey, and related questionnaire, included 'relative importance factors' affecting the choice of flag, rated on a scale of 1 (important) to 5 (very important). 'Flag preference factors' – national-, second-, or open registry – were rated 20~29 for preference degree 'very low', and 90~100 for preference degree 'very high'. The 12 judgement factors, on which the three alternatives were assessed, were:

- U_1 = Crew costs savings.
- U_2 = Costs of meeting maintenance and safety requirements/bureaucratic control.
- U_3 = Capital, insurance and other costs.
- U_4 = Ease of obtaining bank finance.
- U_5 = Fiscal advantages.
- U_6 = Labour quality and availability.
- U_7 = Vessel characteristics (age, size and type, etc.).
- U_8 = Trading region of the world.
- U_9 = Public relations.
- U_{10} = Country-specific comparative advantages (subsidies, economic power and structure, etc.).
- U_{11} = Political considerations.
- U_{12} = Union considerations/recognition.

Subsequently, Step 2 defines the weightings of the judgement factors and a corresponding fuzzy weightings vector P. Based on questionnaire results, the weightings of the twelve factors are:

3.917, 3.750, 1.600, 2.250, 3.000, 3.500, 2.800, 2.900, 2.350, 3.000, 3.300, 3.300

Weightings must satisfy the normalization requirement:

$$0.1098+0.1052+0.0449+0.0631+0.0841+0.0981+0.0785+0.0813+0.0659+$$
$$0.0841+0.0925+0.0925=1 \quad (1)$$

The fuzzy weightings vector P is simply:

$$P = [p_1, p_2, p_3, p_4, p_5, p_6, p_7, p_8, p_9, p_{10}, p_{11}, p_{12}]$$
$$= [0.1098, 0.1052, 0.0449, 0.0631, 0.0841, 0.0981, 0.0785, 0.0813,$$
$$0.0659, 0.0841, 0.0925, 0.0925] \quad (2)$$

Step 3 defines the set of comment factors (linguistic variable X and linguistic value X_i). The linguistic variable refers to the preference degree of flag alternatives, which consist of 'very high', 'quite high', 'rather high', 'high', 'low', 'rather low', 'quite low', 'very low'. According to the importance of the judgement factors, as perceived by the respondents, each factor is given a linguistic value ranging from 20~100. These appear in Table 5.3.

Step 4 constructs the membership function $\mu(x)$. Membership functions are at the core of fuzzy models. The membership function is considered to be both the strongest and also the weakest point of fuzzy set theory (Mansur, 1995). It is strongest because a membership function defines a soft threshold, which allows a smooth and practical assessment of the contribution of the judgement factors to the decision of flag selection, in contrast to a characteristic function, which defines a hard threshold in classical set theory. However, it is also the weakest, because the membership function is regarded as too subjective in relation to its construction. A membership function can be either discrete or continuous. In most cases,

Table 5.3 Linguistic values of judgement factors

Judgement factors	National registry (X_1)	Second registry (X_2)	Open registry (X_3)
U_1	45.75	53.00	72.50
U_2	56.25	52.50	53.75
U_3	60.50	65.00	81.75
U_4	53.75	51.25	56.25
U_5	54.50	58.00	70.00
U_6	72.00	57.50	57.50
U_7	52.50	54.25	53.00
U_8	56.25	55.00	61.25
U_9	71.25	61.25	55.00
U_{10}	71.25	60.75	49.00
U_{11}	72.00	58.25	56.50
U_{12}	55.25	50.75	61.25

membership functions assume continuous forms. There are several alternatives of the functional form: triangular, trapezoidal, Gaussian, bell and sigmoidal membership functions are the most commonly used (Dubois and Prade, 2000).

Several studies discuss empirical methods to construct a membership function based on expert knowledge. With regard to the latter, a number of aspects must be considered for the practical application of fuzzy models to the flag decision (Cornelissen et al., 2000). These include the necessary qualifications of experts, the proper elicitation of expert knowledge for the construction of the membership function, and the methods to test the reliability of membership functions. Reliability is also important with regard to verification and validation of the fuzzy model.

In the case of CFE, membership is given in Gaussian form (Liu, 1998) by:

$$\mu(x) = e^{-0.000278(x-100)2} \quad (0 \le x \le 100) \tag{3}$$

Step 5 computes the degree of membership (r_{ij}) and the fuzzy matrix $R_i = \{r_{ij}\}$. Based on the membership function (3), the degree of membership for the three registry alternatives can be calculated as follows:

$\{r_{1j}\} = (r_{11}, r_{12}, r_{13}) = (0.4412\ 0.5411\ 0.8104)$
$\{r_{2j}\} = (r_{21}, r_{22}, r_{23}) = (0.5874\ 0.5341\ 0.5518)$
$\{r_{3j}\} = (r_{31}, r_{32}, r_{33}) = (0.6481\ 0.7114\ 0.9116)$
$\{r_{4j}\} = (r_{41}, r_{42}, r_{43}) = (0.5518\ 0.5165\ 0.5874)$
$\{r_{5j}\} = (r_{51}, r_{52}, r_{53}) = (0.5624\ 0.6124\ 0.7786)$
$\{r_{6j}\} = (r_{61}, r_{62}, r_{63}) = (0.8042\ 0.6052\ 0.6052)$
$\{r_{7j}\} = (r_{71}, r_{72}, r_{73}) = (0.5341\ 0.5589\ 0.5411)$
$\{r_{8j}\} = (r_{81}, r_{82}, r_{83}) = (0.5874\ 0.5695\ 0.6587)$
$\{r_{9j}\} = (r_{91}, r_{92}, r_{93}) = (0.7947\ 0.6587\ 0.5695)$
$\{r_{10j}\} = (r_{101}, r_{102}, r_{103}) = (0.7947\ 0.6516\ 0.4853)$
$\{r_{11j}\} = (r_{111}, r_{112}, r_{113}) = (0.8042\ 0.6160\ 0.5909)$
$\{r_{12j}\} = (r_{121}, r_{122}, r_{123}) = (0.5731\ 0.5095\ 0.6587)$

Then, the fuzzy matrix is:

$$R_i = \begin{pmatrix} 0.4412 & 0.5411 & 0.8104 \\ 0.5874 & 0.5341 & 0.5518 \\ 0.6481 & 0.7114 & 0.9116 \\ 0.5518 & 0.5165 & 0.5874 \\ 0.5624 & 0.6124 & 0.7786 \\ 0.8042 & 0.6052 & 0.6052 \end{pmatrix}$$

Step 6 determines the fuzzy judgement $S_i = P_i \bullet R_i$ so as to assess the effects of the selected flag.

The fuzzy judgement is:

$S_i = P_i \bullet R_i$ = [0.1098, 0.1052, 0.0449, 0.0631, 0.0841, 0.0981, 0.0785, 0.0813, 0.0659, 0.0841, 0.0925, 0.0925] $\bullet R_i$ = [0.6366, 0.5835, 0.6398]

...and, as a result, alternative 2, i.e. 'second registry', is the optimal choice for ship registry.

5.5   Reversing Chinese flagging out: government intervention and policy competition

In China, as indeed in many other countries throughout the world, international shipping furnishes benefits to the economy that go far beyond the short-term commercial results of the shipping companies themselves. Unfortunately, despite a number of successful policy examples in many countries – especially in the European Union – China has so far failed to realize that the provision of an attractive operational environment to shipping companies, and the resulting improvement in the competitiveness of its flag, can generate multiple benefits to the state that would significantly outweigh any costs (or foregone revenues) involved in the provision of such an environment.

The EU experience: policy competition and the attractiveness of national flags

In most EU countries, the national register drain is continuing, and this development is assuming alarming dimensions with painful consequences. To maintain a large and strong national fleet, most governments of traditional maritime countries have modified their policies moving closer to the situation created by the legislation of open registers. The FOC legislation is often used as a benchmark against which to measure the effects of policies of traditional maritime countries (Veenstra and Bergantino, 2000). In this way, 'policy competition' is being developed between traditional maritime countries and open registers.

Usually, policy competition for attracting and retaining national fleets falls into two main categories (Raines and Brown, 1999): one is an 'incentive-based' approach, aiming to influence flag choice directly by such things as flag preference/discrimination (including cargo reservation); exclusion of foreign flags (cabotage, bilateralism, multilateralism); port surcharges/discriminatory fees; and maritime subsidies such as operating and building subsidies, investment/modernization grants and tax benefits.

The other category consists of a 'rule-based' approach, having a more indirect impact on flag choice by affecting the regulatory and operating environment of a shipowner. Examples comprise various special regimes with particular rules and regulations putting national fleets on more equal footing with those of other countries (Sletmo and Holste, 1993).

The first approach has been used frequently in the past, but it has been unable to prevent the decline of national-flag fleets. The second policy option became increasingly popular during the late 1980s and one of its manifestations is the establishment of what has come to be known as 'international registries': policy solutions aiming to reconcile private profitability considerations and national economic welfare.

At the level of the individual firm, the flag decision should be viewed as similar to any other strategic decision of the profit-maximizing firm and should therefore be taken solely on commercial criteria. On the other hand, every national economy has to maximize its benefits from shipping. When evaluating the economic and social effects of flagging out, governmental authorities ought to consider its welfare effects on the overall economy. Policy solutions should derive from such an evaluation. To this end, the second policy approach represents, at least theoretically, the best of both worlds by combining the advantages of FOCs with those of the traditional maritime countries. International registries should thus be viewed as the point where both the private interests of shipping companies and the wider ones of the national economy are reconciled.

Most EU countries have long-standing maritime traditions and in the last decade the EU has developed a comprehensive and active approach to maritime affairs. By the 1990s, European 'traditional' registers and seafaring employment had declined so precipitously that the Union had to reappraise its maritime strategy. The EU's response to the decline of European shipping, in the face of international structural change in the industry, was set out in its paper entitled 'Towards a New Maritime Strategy' (Commission of the European Communities, 1996).[3]

One of the main objectives of the revised policy was to ensure, through a Community framework for enhancing shipping competitiveness, that ships are 'preferably registered in EU Member States with Community nationals employed on board'. The competitiveness framework encompassed policies on training and employment, research and development, and state aids. As to the latter, the means by which member states could intervene in the market to encourage EU ship registration and employment were defined in the Commission's maritime state aid guidelines (Community Guidelines on State Aid to Maritime Transport). The guidelines establish a more liberal regime for shipping than in any other sector in the EU economy (Department of the Environment, Transport and the Regions, 1998).

Examples of state aids in shipping can be found in most traditional maritime nations of Europe, from the early Norwegian initiative to the German package approved only in 1998. In these initiatives, common elements are fiscal relief measures such as tonnage-based corporate taxation, and the exemption of social charges for seafarers. A notable initiative is the comprehensive package of measures introduced by the Netherlands which, as the Dutch authorities claim, has succeeded in establishing the circumstances in which their maritime industry is now enjoying an economic recovery. The new approach of The Netherlands focuses on creating an attractive business and investment climate in which shipping is seen as the core of the country's maritime business cluster. The central element of the Dutch maritime initiative, reflecting the new focus on Dutch ownership rather than on the Dutch-flagged fleet, is an optional tonnage-based tax regime. The new policy was introduced at the beginning of 1996. Results include a 25 per cent increase in the Dutch merchant fleet, full employment of Dutch seafarers, a renewal of shipbuilding in local yards and the return or relocation to The Netherlands of some forty shipowning or ship management companies (ibid.).

A proposed shipping policy for China to change the situation of flagging out

As noted above, policy competition and government intervention are becoming worldwide instruments to attract and retain national fleets. In this light, China also needs to retain a strong Chinese presence in shipping and, where possible, a strong Chinese-registered fleet to ensure the development of the national economy and security. To this end, China should use the experience of developed maritime countries as a reference. This means that China will have to adjust its shipping policy, making it more appealing to Chinese shipowners and encourage them to register their vessels domestically.

Zhang (1998) quantified and compared the degree of protectiveness (or openness) of shipping policy for typical maritime countries by an integrated Delphi and comprehensive fuzzy evaluation method. According to his results, as shown in Table 5.4, the present protective degree of China's shipping policy is not only lower than that of the USA, but also lower than that of traditional open economies such as France and South Korea. This is consistent with China's general policy of economic openness and its aspirations within the WTO. The country has already undertaken a series of important liberalization initiatives in shipping, including matters such as the abolition of preferential treatment for Chinese companies, cargo reservation, and favourable interest rates for ship finance. However, as far as the competitiveness and endurance of Chinese shipping companies are concerned, Zhang argued that the degree of openness of the Chinese shipping policy is over-advanced.

However, despite the openness of its economy, China is still a developing country requiring appropriate non-discriminatory policies in order to boost its international competitiveness, shipping sector included. With regard to the latter, government policy should focus its efforts on expanding those activities that create the highest value added (Veenstra and Bergantino, 2000). In certain traditional maritime countries, such as The Netherlands, shore-based maritime activities, most prominently

Table 5.4 Comparing the protective degree of shipping policy between typical maritime countries and China

Mean degree of protectiveness	USA			France			South Korea			China		
	1982	*1992*	*2000**	*1982*	*1992*	*2000**	*1982*	*1992*	*2000**	*1988*	*1994*	*2000**
$\bar{P}$	0.85	0.71	0.62	0.81	0.56	0.16	0.75	0.59	0.25	0.88	0.14	0.59

*Note: Expected for typical maritime countries and suggested for China.
Source: Journal of Wuhan Transportation University (China), December 1998.

shipping management, contribute most of the value added generated by the shipping industry. This is not the case in China, however, where activities such as multimodal transport, ship broking, insurance, etc. are far less developed. Shipping value added in China is generated by the ship itself and thus any shipping policy should be geared towards attracting and retaining the national fleet.

Shipping policy in China, as in most other countries, should aim to level the playing field in international competition through such measures as fiscal relief; shipbuilding terms similar to those offered to foreign buyers; the abolition or lowering of ship import tariffs and value added taxes on international trades; waivers on social charges for seafarers; tonnage-based tax regime, etc. In addition, the government should increase investments in maritime education and training in order to encourage the development of a skilled and flexible maritime labour force.

Finally, the possibility of establishing an international ship register, as suggested in this chapter, should be considered seriously, especially in order to attract back those vessels which have already been registered overseas.

As mentioned above, 'international', 'parallel', or 'offshore' ship registers, as they are often called, are widely adopted in Europe. Their main objective is to level operating costs with those prevailing under FOC registration, while, simultaneously, maintaining high technical standards and effective implementation of international conventions. By also flying the national flag, ships in the parallel register enjoy a high reputation amongst shippers and charterers.

In the case of China, such a register should have the following characteristics: the type of vessels registered should be Chinese-owned (including domestic and FOC); the manning system should be eased and the required number of national seafarers reduced; and the taxation policy should be streamlined to bring it in line with the tonnage tax policies of other countries or FOCs. Such tax policies may include exemptions from tonnage dues, annual tax, and income tax by seafarers.

5.6   Conclusions

In China, as in most other countries, flagging out has been shown to have serious negative impacts on China's shipping development and national economy. To change the situation, China has to adjust its shipping policy in the light of the experiences in other countries that have faced this unfavourable development. Such policy reorientation should be based on the evaluation of the economic, social and political effects of shipping registry alternatives. China should adopt more preferential shipping policies such as favourable shipbuilding arrangements; tax exemptions for ships in international trades; waivers of social charges on seafarers; tonnage-based

corporate taxation; and greater support for maritime education and training in order to maintain skills and a flexible labour force. Finally, the establishment of a parallel register, along the lines of the European experience,
as suggested in this chapter, would help in attracting back Chinese-owned
vessels.

Notes

1 This chapter was previously published as Haralambides and Yang (2003). The
   editors are grateful to Elsevier for granting permission to reproduce it herein.
2 The authors gratefully acknowledge financial support from the China Holland
   Education and Research Centre (CHERC).
3 The then EU Transport Commissioner Neil Kinnock described the paper as
   'A Green Paper with White Edges'. The paper was drafted by Kinnock's Core
   Group on Maritime Transport – one of the authors was a member of that group –
   and subsequently debated at the Roundtable Conference of Erasmus University
   Rotterdam in October 1996. It is considered as one of the most important shipping documents ever drafted by the EC and it is recommended reading for every
   student of shipping policy.

References

Bergantino, A.S. and O'Sullivan, P. (1999) Flagging Out and International Registries:
   Main Development and Policy Issues, *International Journal of Transport Economics*,
   26, 446–71.
Commission of the European Communities (1996) *Towards a New Maritime Strategy*,
   COM(96) 81 final, Luxembourg: Office for Official Publications of the European
   Communities.
Cornelissen, A.M.G., Van den Berg, J., Koops, W.J., Grossman, M. and Udo, H.M.J.
   (2000) Assessment of Sustainable Development: A Novel Approach Using Fuzzy Set
   Theory, Erasmus University, Rotterdam.
Department of the Environment, Transport and the Regions (1998) *British Shipping:
   Charting a New Course*, London: DETR.
Dubois, D. and Prade, H. (2000) *Fundamentals of Fuzzy Sets*, Boston: Kluwer Academic
   Publishers.
Haralambides, H.E. and Yang, J. (2003) A Fuzzy Set Theory Approach to Flagging
   Out: Towards a New Chinese Shipping Policy, *Marine Policy*, 27(1), 13–22.
Institute of Shipping Economics and Logistics (2000) *Market Analysis 2000: Ownership
   Patterns of the World Merchant Fleet*, Bremen: ISL.
Liu, S. (1998) *Systematic Engineering Theory on Transport*, Beijing: The People's
   Transport Publishing House.
Mansur, Y.M. (1995) *Fuzzy Sets and Economics: Application of Mathematics to
   Non-Cooperative Oligopoly*, Aldershot: Edward Elgar Publishing.
Ministry of the Communication (MoC) (1999) *China Shipping Development Annual
   Report 1998*, Beijing: The People's Publishing House.
Raines, P. and Brown, R. (1999) *Policy Competition and Foreign Direct Investment in
   Europe*, Aldershot: Ashgate.
Saaty, T.L. (1980) *The Analytic Hierarchy Process: Planning, Priority Setting, Resource
   Allocation*, New York: McGraw-Hill.

Sletmo, G.K. and Holste, S. (1993) Shipping and the Comparative Advantages of Nations: The Role of International Shipping Registers, *Maritime Policy and Management*, 20, 243–55.

United Nations Conference on Trade and Development (UNCTAD) (1999) *Review of Maritime Transport*, New York: United Nations.

Veenstra, A.W. and Bergantino, A.S. (2000) Changing Ownership Structures in the Dutch Fleet, *Maritime Policy and Management*, 27, 175–89.

Zhang, P. (1998) Optimal Options on our Civil International Shipping Economic Policy, *Journal of Wuhan Transportation University*, 22, 651–5.

6

Deregulation, Competitive Pressures and the Emergence of Intermodalism

Sophia Everett

6.1 Deregulation, competitive pressures and the emergence of intermodalism

Over the course of the last decade or so, deregulation has dramatically altered the face of Australian industry and associated services. In the transport sector, in particular, changes have been dramatic and deregulation has led to pervasive changes in market structure, to the actual ownership of infrastructure and to a shift in strategic focus from a public utility to one of commercial viability and market orientation.

The imposition of intense competitive forces as a result of deregulation has meant that traditional public sector organizations such as railways and ports have been transformed into government-owned businesses. Furthermore, it has meant that, in order to survive in the increasingly competitive environment, these organizations have been forced to reinvent themselves. Freight Corp, the NSW rail freight operator, for example, has introduced an intermodal service providing warehousing, trucking operations as well as a shuttle rail link to the port of Sydney-Portlink; and the Brisbane Port Corporation has established the Brisbane Multimodal Terminal – a facility linking land and port interfaces; similarly, stevedoring companies are also providing end-to-end intermodal services including warehousing, rail operations, freight forwarding as well as coastal shipping services.

Deregulation, which has been part of a broader microeconomic policy implemented by government over the last two decades with the objective of creating a more competitive environment, has driven many of these changes. This trend has been in direct contrast to the protectionist policies which had shielded Australian industries from competitive forces in the past, but which could no longer be sustained in an increasingly competitive global environment.

Deregulation has had a particular impact on the public sector. Not only did it open the market to private sector competitive forces, but in so doing it eliminated previously held monopoly positions. Essential transport infra-structure in Australia, such as ports and railways, had traditionally been developed and continue to be operated by state governments – they were public utilities developed by the state in order to enhance economic growth and trade and their *modus operandi* reflected this. In the area of rail, for example, funding strategies up to the 1970s reflected prevailing views that railways were essentially public utilities operating on noncommercial grounds, although attempts at transforming them into profitable businesses had periodically been undertaken with varying degrees of success. But within an emerging, more competitive economic climate, growing deficits, which had reached a peak of $2.7 billion Australia-wide by 1983/84,[1] could no longer be sustained by state governments. As a result, dramatic change was imminent.

This chapter examines briefly the policy of deregulation as an element of microeconomic reform in Australia and discusses the impact of deregula-tion on the provision of transport services and the emerging intermodal environment.

6.2   Microeconomic reform

Microeconomic reform in Australia has been concerned with enhancing efficiency and competitiveness by exposing Australian industries, within both the public and the private sectors, to competitive forces. It has been implemented through a number of steps.

Labour market reform

Initially microeconomic reform focused only on some aspects of the labour market.[2] This was driven by a relatively simplistic notion that competitive positions could be established through a downsizing of the labour force. In the area of Australian flag shipping, crew numbers were reduced from some seventy or eighty to approximately forty-five per vessel; in ports the numbers working in container terminals were reduced from 6,000 to approximately 3,000; in the area of rail the numbers in Queensland Rail were reduced from 24,000 to some 14,000 at the present time with further reductions imminent; and port authorities also reduced their numbers dra-matically – the numbers employed by the former Maritime Services Board of NSW were reduced from 3,000 in the mid-1980s, when the commercial-ization process was instigated, to some 600 in 1995 when NSW ports were corporatized.

While downsizing and labour market reform led to improvements in pro-ductivity it did not create a competitive position per se. It led to improve-ments in technical and operational efficiency which is a necessary but not a sufficient condition to capture a competitive position.

Public sector reform

Microeconomic reform initially focused on private sector interests such as the automobile, iron and steel industries. It was subsequently to focus on the public sector, with arguably more dramatic effects. This was particularly the case post-Hilmer[3] and the formal adoption by government of National Competition Policy.

The thrust of Hilmer and public sector reforms was to make government businesses competitive – indeed, similar to their private sector counterparts. One way of achieving this was by market deregulation and by commercializing and/or corporatizing government instrumentalities – those engaged in commercial operations. The expectation was that this restructuring would lead to government-owned businesses emulating private sector company practices with the aim of maximizing both efficiency and profit.

The outcome of public sector reform led to ownership and corporate changes in what had traditionally been government-owned operations – corporatization and privatization became the norm. The two Australian flag carriers, ANL and State ships, were privatized; the Victorian bulk ports, Geelong and Portland, were privatized along with the management of the port of Hastings. Most other Australian ports, with the exception of those in Western Australia, were corporatized.[4]

A fundamental aim of reforming the public sector was the establishment of a 'level playing field' – that is, removing any perceived advantages associated with public sector ownership. This meant the universal application of the rules of Part IV of the Trade Practices Act to all business activities in Australia – private and public sectors alike – and the removal of immunity associated with common law doctrine, which the public sector had traditionally enjoyed.

Deregulation

Another crucial element of public sector reform has been deregulation. This means that monopoly positions previously held by rail operators and port authorities,[5] for example, were removed, at least technically. Deregulation in the rail sector has meant that rail operators are no longer restricted by state boundaries but can now operate interstate. Furthermore, deregulation in the rail sector has meant the separation of below-track and above-track operations – that is, the separation between the ownership of the track infrastructure and the user of that infrastructure. This has been by way of introducing 'open access' regimes, which means that competition now exists between users of that track.

Deregulation has, as a result, created a highly competitive market with the entry of numerous operators. Competition has come not only from former public sector rail operations, now either corporatized or privatized, but from a spate of non-traditional rail operators – former trucking companies and freight forwarders as well as stevedores, for example.

6.3　The mechanism for change

Deregulation has been, *inter alia*, about enhancing efficiency and competitiveness by exposing Australian industries – public and private sector alike – to competitive forces. In effectively implementing this policy, particular sensitivities had to be overcome between governments at the federal and state levels before deregulatory reform could be introduced. Transport infrastructure fell within the jurisdiction of state government control and any attempt to intervene by the federal government had been vehemently opposed in the past by state governments. Microeconomic reform, however, of which deregulation is an important component and which was to culminate in a national competition policy, was initially a federal government initiative. As infrastructure ownership lay within state government jurisdiction, however, effective implementation required the cooperation of state governments – not a straightforward matter given the nature of Australian politics.

Agreement was reached between the federal and state governments that a cooperative and consensus approach would be adopted and a joint Commonwealth–State committee on microeconomic reform was established whose aim was 'to foster the development of a common national market for goods and services'.[6]

Reform, which was to culminate in a formal Competition Policy, was pursued further with the appointment by the Prime Minister in 1992 of the National Competition Policy Review Committee (the Hilmer Committee). Fundamental principles of competition policy included that:

- No participant in the market should be able to engage in anti-competitive conduct against the public interest.
- As far as possible, universal and uniformly applied rules of market conduct should apply to all market participants regardless of the form of business ownership.
- Conduct with anti-competitive potential said to be in the public interest should be assessed by an appropriate transparent assessment process, with provision for review, to demonstrate the nature and incidence of the public costs and benefits claimed.
- Any changes in the coverage or nature of competition policy should be consistent with, and support, the general thrust of reforms:
  - To develop an open, integrated domestic market for goods and services by removing unnecessary barriers to trade and competition.
  - In recognition of the increasingly national operation of markets, to reduce complexity and administrative duplication.[7]

The Hilmer Report gave a good deal of attention to the need for competition in the private sector but, as Quiggin (1996) points out,[8] the subsequent policy process was dominated by initiatives directed at the public sector. In particular, the thrust of the Hilmer reforms was to make government businesses competitive and like their private sector counterparts. One way of achieving this was by market deregulation and by corporatizing government instrumentalities engaged in producing goods and services. The expectation was that this restructuring would lead them to operate like private sector companies with the aim of maximizing profit although they would continue to have responsibility for community service obligations.[9]

The aim of reforming the public sector was also about the establishment of a 'level playing field'. This involved extending an identical regulatory regime to the private and public sectors alike. This meant that the rules of Part IV of the Trade Practices Act would apply universally to all business activities in Australia[10] – private and public sector alike – and the removal of immunity associated with common law doctrine, which the public sector had traditionally enjoyed.[11]

An important feature of deregulation was the separation of government businesses into discrete units. This structural separation occurred in three main areas:[12]

- The separation of regulatory and commercial functions.
- The separation of natural monopoly and potential competitive activities; and
- The separation of potential competitive activities.

Horizontal divestiture involved the division of a single enterprise into several smaller enterprises. In effect in the NSW rail sector this meant that a separation occurred between the owners of the below-track infrastructure from the actual operators. This effectively eliminated the monopoly position of the operator, as track infrastructure would be available to other players thereby introducing competition in the market.

In August 1993 the Hilmer Report was submitted for joint agreement by the state and federal governments and in 1995 the federal, state and territory governments signed the Competition Principles Agreement and the Conduct Code. Further constitutional obstacles had to be overcome as this Agreement concerned price oversight but, as King and Maddock[13] point out, under the Australian Constitution the commonwealth government does not have the constitutional power to control prices charged by state-owned enterprises. In order to overcome this hurdle, it was agreed that an independent price review body for all areas of government businesses, whether state or federal, would be established with the responsibility for price supervision in all areas of government-owned businesses which are monopolies or near monopolies and to ensure that monopoly powers were not abused – the Australian Competition and Consumer Commission.

Furthermore, under the Agreement the notion of 'competitive neutrality' was introduced. This meant the establishment of a level playing field and that public sector businesses would not have competitive advantage by virtue of ownership. That is, public sector businesses would operate like any private sector business – they would be subject to the same tax obliga- tions, interest rates and social and economic regulations as their private sector counterparts.

A crucial element of the Competition Principles Agreement of particu- lar significance to the rail sector was the provision of third party access to essential infrastructure. The Agreement stipulated that owners of infrastructure facilities, including below-track rail infrastructure, would be required to provide third party access to those facilities. This removed the monopoly power of state government-owned railways. While the monopoly position of below-track infrastructure owners was maintained, third party access meant that the actual use of the infrastructure was accessible to other operators – the monopoly position of the rail operator in effect disintegrated.

6.4 Impact on the transport sector and the emergence of inter-modalism

The impact of deregulation has been so significant that stevedoring compa- nies and rail and truck operators as traditional modal operators have significantly restructured their operations in order to become market- focused third-party service providers of a range of integrated functions. No longer does a stevedoring company, such as Patrick Stevedores, merely handle ships. It is now in the process of providing warehousing, has formed an alliance with a national rail operator to provide a regular rail service between Adelaide and Melbourne, is in the business of freight forwarding and has purchased a coastal shipping operation. It is also con- sidered a likely contender for the soon-to-be-privatized National Rail Corporation and Freight Corp. In short, a stevedoring company is now in the business of providing a service end to end. Similarly, a traditional trucking company, such as Toll Holdings, has extended its activities towards end-to-end control by purchasing one of the privatized Victorian ports, Geelong. It provides, in addition, rail services Australia-wide as well as warehousing – in short, Toll Holdings is also focusing on capturing the market end to end.

This development of providing an end-to-end service in one sense is not new – the development of containerization during the late 1960s was to introduce an end-to-end delivery system. The strategy was not imple- mented at the time, however, for a number of reasons including institu- tional barriers but primarily because the commercial pressures did ₁ exist.[14] By the end of the 1990s, however, the market is quite different that existing during the 1970s with prevailing competitive pressures on the traditional transport operator to the extent that endeavours to capture or retain market share are essential if the company is to survive. Intermodalism as we are experiencing it, Robinson[15] argues, is about the integration not only of a number of modal operations, but about integrating a range of operational functions and of integrating a package of functions which gives effective control over part or all of the freight movement system. The intermodal operator is strategically positioned to value add and capture market share by delivering an efficient and effective end-to-end product to the customer.

What is new about the current developments is that intermodalism is a critical aspect of delivering value and therefore creating competitive advantage and of expanding market share by providing a customer-focused end-to-end service. The introduction of intermodalism is in response to an increasingly competitive national and international trading environment. It is driven by the belief that efficiency within the transport system is created when a single operator, or an alliance of operators, has control of the end-to-end operation with the ability of eliminating bottlenecks within a system likely to occur when a multitude of operators are involved in a highly segmented operation.

Deregulation has been the enabling factor in the development of an intermodal system. Rail operations are no longer the prerogative or responsibility of state government operators constrained by state boundaries and providing essentially line haul services. The intermodal operator now routinely integrates a multitude of transport-related functions – from warehousing to road and rail services, port functions and, in some instances in the Australian context, that of coastal shipping. Deregulation has meant that rail operations are no longer the exclusive prerogative of state government operators and the provision of open access regimes has meant that other operators have entered into the rail market in competition with the former state government monopolists. In effect, deregulation has meant that new entrants into the Australian rail freight market are either niche railway operators[16] or are established rail operators with the ability to access inter-state markets.[17] In addition, deregulation has enabled a host of third party service providers, Patrick Stevedores and Toll Holdings, for example, who are able to take advantage of a deregulated port and rail market to expand its existing operation to form part of its end-to-end intermodal service seriously threatening the incumbent's market share and, for the

cing competition into the market.

ction discusses the way in which Freight Corp, the NSW
t corporation, responded to deregulation and exposure
ve forces and redesigned its port-linked rail container

6.5   Port Link: a case study

The newly deregulated rail market has meant that incumbent operators, corporatized public sector rail operators, such as FreightCorp and Queensland Rail, must compete with new entrants in the market – with traditional trucking companies and stevedoring operators who are using access to rail to provide an end-to-end intermodal service and who tend to enjoy competitive advantage by virtue of the fact that they have lower overheads and capital investment in infrastructure. This has created a new trading environment and traditional rail operators are themselves developing intermodal strategies in order to either retain existing or capture additional market share. Both Freight Corp and Queensland Rail are developing intermodal strategies in a bid to expand their businesses which, until recently, have essentially involved the haulage of bulk products on a line haul basis.

Freight Corp is implementing this strategy through its Port Link pro-gramme – a concept based on the recognition that efficiency in transport comes from eliminating bottlenecks and smoothing interfaces – an efficiency derived from the provision of an end-to-end intermodal trans-port service from producer to the port with the integration of road and rail operations, intermodal terminal, warehousing and port services. Freight Corp provides these services not by owning all essential infrastructures but by joint venturing with other modal operators. Figure 6.1 illustrates the Port Link concept comprising a series of inland terminals or centralization centres integrated with rail and road links to the port.

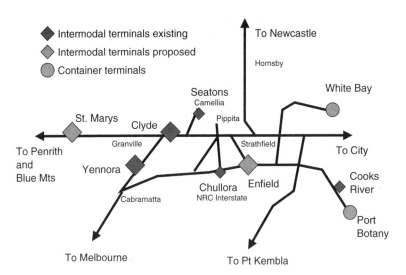

Figure 6.1   Metropolitan intermodal terminals – Sydney

The push into intermodalism in Australia differs in some aspects with that experienced overseas where large shipping companies, such as Maersk, have ventured into ownership of much of the requisite infrastructure. Maersk as a result owns not only the shipping fleet but also port terminals, road and rail operations and inland distribution centres. In the Australian experience where distances are extensive with relatively low volumes the trend has been to form alliances and/or joint ventures rather than to opt for outright ownership.

Freight Corp, for example, has formed alliances with road operators who provide the road leg to the intermodal facility, which is linked by rail with the port. In some markets, Freight Corp concentrates on providing the line haul leg and while owning the intermodal terminals has contracted the management and operation of these facilities to the trucking companies. In other instances the Freight Corp contribution to the alliance may be its expertise as a line haul operator and the supply of equipment such as gantry cranes and forklifts.

Port Link at the present time does not include a blue water component, though agreement has been reached with a number of shipping companies, which has led to Freight Corp controlling the movement of containers, by road and rail to and from the port on behalf of the shipping companies. Maersk, Mediterranean Shipping and Fesco, for example, have established links with Freight Corp for services enabling them to market and sell 'through' rates with Freight Corp providing the land leg. Other services offered by Freight Corp to the shipping companies involve the shuttle concept between metropolitan intermodal terminals at Clyde and Yennora thereby avoiding congestion at the port.

At the port end, cooperation with the stevedores, P&O and Patrick, has reduced the bottlenecks and smoothed the interfaces between the rail terminal and the port end of the chain. The development of Port Link has meant an increase in rail receival of some 17 per cent – from 80,000 TEUs in 1997 to approximately 100,000 TEUs in 2000. This growth in itself initially created a bottleneck at the rail/terminal interface associated with the irregular arrival patterns of export containers; availability of wagons for import deliveries; and conflicting demand on resources at the wharf. Cooperation between Freight Corp and the stevedoring companies, however, has led to a time scheduling for entry into the terminal smoothing that bottleneck. The stevedores now provide slot times for Freight Corp to shunt its rakes into position in the terminals with wagons being stripped and back loaded in a prescribed time frame. The Port Link concept is based on a new paradigm in the transport industry. Rather than segmentation within a transport chain, intermodalism is based on the notion of integrating a number of modes. It is about control of the supply chain either by way of owning infrastructure, as has occurred with large shipping companies such as Maersk, or through pursuing joint venturing, as is provided by

Port Link. It has been driven by fierce competition within the market as a result of deregulation and the need for value adding in order to capture or retain market share.

6.6   Concluding comments

Deregulation within the Australian transport sector has had a number of profound impacts, which have altered dramatically the structure and nature of the industry. It has increased the number of players in the market; it has introduced a different type of player providing a different type of service – one from a traditional line haul or road operation to an intermodal end-to-end third party service provider. Deregulation has meant that the monopoly positions held by former government rail and port operators have been removed. In the rail sector, deregulation and access to rail infrastructure has meant that any approved operator can now engage in the business of running trains – but it is a business with a difference – one that is part of an intermodal operation. These developments have created a highly competitive market and in those instances where actual competition has not been introduced, deregulation has made the market contestable. This, in itself, has led to competitive outcomes. Rail freight rates for the cartage of coal in the Hunter Valley of NSW, for example, have been reduced by some 25 per cent as the result of competitive pressures within that market.

A major impact of deregulation has been that transport operators have been forced to restructure and refocus and, in the face of growing competition, have been forced to reinvent themselves. Rail operators such as Freight Corp are no longer simple line haul operators in urban container movement markets, for example; rather, they have become market-focused third party service providers of a range of integrated functions.

Notes

1   House of Representatives Standing Committee on Communications, Transport and Microeconomic Reform (1998) *Tracking Australia*, July p. 109.
2   Some attempts to introduce change to work practices was also attempted – multiskilling on Australian flag shipping, for example.
3   Report by the Independent Committee of Inquiry (1993) *National Competition Policy*, August.
4   The Western Australian government adopted a commercialization model. In effect, this was not unlike the corporatization model adopted in other states but it lacked the associated legislative changes.
5   Competition between port authorities is rather limited in virtually all Australian ports as, with very few exceptions, commercial operations within a port are undertaken by the private sector.
6   Independent Committee of Enquiry (1983: 37).
7   Ibid., p. xviii.

8   J. Quiggin (1996) *Great Expectations – Microeconomic Reform and Australia*, Sydney: Allen & Unwin, p. 187.
9   The significance of this policy was that while government businesses would continue to have the responsibility for carrying out social responsibilities of government, they would in future be funded by Treasury separately rather than being cross-subsidized by other commercial businesses as had been the case in the past.
10   Independent Committee of Inquiry (1993, p. xxi).
11   Quiggin, *Great Expectations*, p. 190.
12   Independent Committee of Inquiry (1993, p.217).
13   King and Maddock (1996) *Unlocking the Infrastructure*, p. 40.
14   Robinson and Everett (1998).
15   Robinson (1998).
16   Including Austrac, Silverton Railways, Lachlan Valley Rail.
17   FreightAustralia (the privatized arm of the former VicRail) NRC, QR, WestRail and FreightCorp.

References

House of Representatives Standing Committee on Communications (1998) Transport and Microeconomic Reform, *Tracking Australia*, July.
King, J. and Maddock, G. (1996) *Unlocking the Infrastructure*, Sydney: Allen & Unwin.
Quiggin, J. (1996) *Great Expectations – Microeconomic Reform and Australia*, Sydney: Allen & Unwin.
Report by the Independent Committee of Inquiry (1993) *National Competition Policy*, August.
Robinson, R. (1998) *Intermodal Systems Management*, Centre for Maritime and Intermodal Systems MGSM, Macquarie University.
Robinson, R. and Everett, S. (1998) *Managing Intermodal Systems Center*, Maritime and Intermodal Systems MGSM Macquarie University.

7
Good Governance and Ports as Tools of Economic Development: Are they Compatible?

Mary R. Brooks

7.1 Background

Over the past two decades, there has been a widespread trend towards the devolution of government-owned entities such as ports. Such devolution of responsibility to the private sector has been undertaken in the belief that the overall level of social welfare will be improved. The Government of Canada began the process of devolution for airports in 1987 and ports were the next logical step. This was accomplished for ports via the implementation of the *National Marine Policy 1995*. Its intention was to secure the benefits of commercially driven business decision making in organizations previously run by governments and, at the same time, to secure compensation for prior taxpayer investments. For any devolution programme to be successful, the government must create appropriate governance structures and processes for the devolved entity.

For many communities, the shift of governance to locally controlled entities is viewed as a progressive policy step to empowering local community economic development. It has been argued (Parr, 1981) that airports and ports are 'growth poles' that can serve as economic catalysts to attract industry to locate and/or invest in their vicinity. For that community, devolution is seen as an opportunity to use the facility as a tool of economic development.

With these two perspectives in mind, this chapter will focus on the principles of good governance, and consider how they are translated into the organizational structure and processes of newly devolved entities. If the government's objective is a commercially driven organization that reimburses the taxpayer for prior investment, and the objective of the community is its own economic development, are the two compatible? Can commercially oriented governance structures work to meet the expectations of both government and the community? This chapter intends to begin the discussion rather than answer the question definitively; that would be the subject of future empirical research.

7.2 What is governance?

Before a meaningful discussion can begin, two terms must be defined: governance and stakeholder. The OECD (1999) defined corporate governance as:

> ...the system by which business corporations are directed and controlled. The corporate governance structure specifies the distribution of rights and responsibilities among the different participants in the corporation, such as the board, managers, shareholders and stakeholders, and spells out the rules and procedures for making decisions on corporate affairs. By doing this, it also provides the structure through which corporate objectives are set, and the means of obtaining those objectives and monitoring performance.

Sternberg (1998: 20) provides a much shorter definition. According to this, corporate governance is concerned with: '...ways of ensuring that corporate actions, assets and agents are directed at achieving corporate objectives established by the corporation's shareholders'.

According to Sternberg (1998), criticisms of corporate governance arise from confusing the term governance with government. She argues that it is a mistake to criticize corporations for failure to achieve public policy objectives or to accord greater importance to their stakeholders, on the assumption that stakeholder participation will provide for better governance.[1] This raises a question: who is a stakeholder?

Stakeholders are groups affected by the decisions of the corporation, e.g., employees, customers, the greater community, advocates for the environment or for product safety. When an organization is devolved, stakeholders raise concerns that their interests will no longer be considered, and that society's interests will be ignored in the pursuit of greater efficiency and profits. It is argued that stakeholders enjoy greater protection when an organization is managed by government than when it is managed by a private sector entity.

In the traditional private sector model, corporate governance is the structure, roles and responsibilities that provide the means by which the organization is managed as an economic entity, based on the objectives of the corporation. In devolution, each devolved entity faces an identity crisis: does it co-opt the objectives of government, or identify its own in keeping with the views of the newly created Board, or co-opt those of community stakeholders? It is the purpose of this section to review basic governance concepts. We will then return to this question closer to the end of the chapter.

Alkhafaji (1989) argues that corporations also have social responsibilities, and that they must respond to the broader issues of society if they are to survive; this is a contentious, even divisive viewpoint. As the focus of this chapter is the governance systems that lead to positive economic development outcomes, it does not intend to explore corporate governance in this broader sense of social responsibility. However, it is proposed that good governance practice requires the imposition of a system of rules and responsibilities compatible with the strategic intent of the organization and its vision of the future. A particular organization may choose to incorporate broader social responsibilities as part of its vision, or may opt to serve only its owners. The driving force is one of strategic intent.

To understand the role of strategic intent, it is important to begin with a review of strategy and structure. Chandler (1962) proposes that structure (of the organization) should reflect the product-market strategy of the firm. That is, that organizational structure follows strategy. He positions strategy as the process of goal identification and the plan for achieving that goal, including the allocation of resources for the implementation of the strategic plan. Galbraith and Kazanjian (1978) believe Chandler's approach to be too narrow and that simply matching strategy and organization structure is not sufficient. They introduce the concept of a compatible configuration of organizational structure, processes, systems, and human resources to achieve effective financial performance. Therefore, the performance of the firm is also a product of industry structure and strategy. Effectiveness of the structure is a function of 'fit', or how consistent or congruent all of the organization's dimensions are with the strategy that is to be pursued. Understanding the firm's core skills is an important component of strategy formulation because fit results if these core skills are matched correctly to strategy.

It is widely argued that competitive advantage accrues to the firm with early strategy–structure fit and that the advantage is eroded over time as other firms seek to gain a better fit. If a firm then changes its strategy, it might therefore change its structure. This does not mean, however, that a firm must change its corporate governance practices. Governance practice, and the principles embodied in that practice, rise above the strategy–structure debate and act as guideposts to assist the organization in realizing its full potential in gaining excellent fit.

This means that governance is a notion that can be applied to more than just corporations. While governance principles are applicable to all relationships between businesses and their shareholders, they are also suitable to relationships between governments and their voters and taxpayers, between public/private agencies and their stakeholders, between organizations and those who establish them to undertake activities on their behalf.

Kaufmann, Kraay and Zoido-Lobaton's (2000) definition of governance reflects this broader applicability:

The traditions and institutions by which authority in a country is exercised for the common good. This includes (i) the process by which those in authority are selected, monitored and replaced, (ii) the capacity of the government to effectively manage its resources and implement sound policies, and (iii) the respect of citizens and the state for the institutions that govern economic and social interactions among them.

This chapter does not intend to explore the corporate governance literature extensively. For that, the reader is referred to Keasey et al. (1999). This four-part, weighty collection of essays on the topic runs approximately 2,300 pages, examines the topic from its history in the days of Adam Smith to appropriate mechanisms for accountability and performance, and draws conclusions from differing disciplines and legal perspectives. There is no shortage of material for debate on corporate governance in the private sector; its content is entirely appropriate and applicable to private sector ports. The dilemma is that there has been little research, either empirical or theoretical, on appropriate governance structures and systems for devolved transport entities, particularly for those cases where profit maximization or full divestiture are not the government's intent. It is also important to distinguish between private and privatized ports, as the latter implies a transition to a new-to-the-organization system of governance.

There is very little literature on appropriate governance practices suitable for devolved ports under non-privatization models (discussed later), although the World Bank (undated) *Port Reform Tool Kit*, it may be argued, is providing a first step. This leads to the conclusion that the literature on principles of good governance for private sector models needs to be examined and then adapted to suit differences between privatization models and other models. This paper uses the Canadian corporate governance code, applicable to Canadian publicly traded companies, to illustrate a potential set of principles of good governance that a devolved transport company might adopt or that a government might incorporate into devolution policies. It then explores how those might be implemented within any organization, before discussing the very limited literature on port devolution governance. Suggestions are made about the compatibility of governance codes with the achievement of economic development objectives a port might have, in order to start the debate about what type of governance research maritime economists should conduct.

7.3　What are the principles of good governance?

Recognizing that good governance practices are essential in delivering value to shareholders, the Toronto Stock Exchange (TSE) developed a set of 14 governance principles in its landmark 1994 study (also known as The Dey Report) on corporate governance practices within publicly traded companies in Canada. The Dey Report was certainly not the first, or the only, investigation into corporate practices; Carver (1990) examined governance in non-profit boards and the Cadbury Commission (1992) explored governance issues in UK publicly traded companies. Subsequent to the release of The Dey Report, there has been a fruitful discussion of governance issues in publicly traded companies (Conference Board of Canada, 1998; OECD, 1999) and in government (World Bank Institute, 2000). The Securities and Exchange Commission in the United States has focused considerable energy on establishing standards of reporting practice and disclosure by boards and management. Codes of Governance, as guidelines of this type are known, have been widely implemented to protect the rights of shareholders (Aguilera and Cuero-Cazurra, 2000), and stakeholder theorists have argued in the strategic management literature that their scope should be broadened to include greater social responsibility content.

This chapter relates primarily to the 14 principles established by the TSE (TSE-14) as voluntary guidelines to boards for no other reason than because this code is the one best known to the author. It is assumed that these principles are broadly applicable to other jurisdictions, but that each jurisdiction will, as part of its due diligence in devolution implementation, examine those it may wish to impose as part of the regulatory environment within which the devolved entity will operate.

The most important word in the above paragraph is 'voluntary'. In developing the TSE-14, the Toronto Stock Exchange recognized that not all guidelines may be deemed appropriate by boards for the particular circumstances of the companies they serve. Suggested practice is to incorporate into the annual report a list of the guidelines and the board's acceptance or rejection of each. This acceptance or rejection must be accompanied by a discussion of actual performance in the case of the former or rationale in the case of the latter. This report is viewed as a means by which shareholders can evaluate the philosophy of management and make an informed decision about investing in the corporation. The TSE-14 principles have been adopted in Canada by many non-profit boards as well, and therefore provide a focal point for the discussion below. Furthermore, there is the added benefit that the TSE conducted a review of corporate practice in Canada five years after the publication of the guidelines (TSE, 1999). The Appendix at the end of this chapter provides a summary of the principles, their rationale and the 1999 findings on the extent of their adoption by Canadian publicly traded companies.

A review of the Appendix at the end of this chapter highlights that relatedness is a key theme embodied in the TSE-14. Basically, an inside director is one who is an officer or employee of the corporation, and therefore is related. Outside directors are usually unrelated, but if they have material interests, provide services to the corporation, or otherwise benefit from the activities of the corporation, they are deemed to be related. A conclusion that should be drawn is that independence of directors from management is a basic tenet of transparent governance, one intended to serve the interests of the shareholders as opposed to the interests of management.

As is evident from the Appendix, for-profit corporations in Canada are having difficulty meeting the ideals established by The Dey Report. The TSE-14 have not been as readily adopted as was hoped. Kazanjian (2000) notes that TSE guideline 5, the requirement to assess board effectiveness as a whole, the effectiveness of its committees and the contribution of individual directors is ranked last in terms of compliance in a survey of more than 600 TSE-listed companies. Less than 20 per cent of corporations had any formal processes in place to evaluate performance on this guideline. Furthermore, benchmarking against the TSE-14 has not happened to the extent anticipated. The Dey Report provides an ideal, as well as a Code of Governance respected by many shareholders, but it has not as yet instilled the discipline that many had expected. It does, however, provide a yardstick for discussion of good governance practice for both private and not-for-profit companies.

As may be obvious, one critical aspect of governance systems is the identification of the role played, or to be played, by each of the parties involved. This is explained in Table 7.1.

As governments proceed to devolve ports from being government-owned and -controlled to more business-like models, it is necessary for them to examine why private sector governance models work; the role to be played by each of the parties needs to be understood. If the new governance model includes elements of the private sector model illustrated in Table 7.1,

Table 7.1　Who does what? Roles of the corporate governance team

Shareholder	Board	Individual Director	Chair of the Board	Board Committee & Committee Chair	CEO
Provider of purpose Provider of capital	Leader Overseer Steward Reporter	In addition to general board roles: Learner Inquirer Influencer	In addition to general board and individual director roles: Achiever of diversity Consensus builder Solidarity promoter	In addition to general and individual director roles: Policy developer Advisor Planner Recommender	In addition to general board roles: Director Initiator Implementor Mentor

Source: Conference Board (1998), part of table 1, page 1.

Table 7.2   Typical responsibilities in for-profit corporations

ACTIVITY	Owner functions	Board functions
Leadership & stewardship	• Elect/appoint directors • Appoint auditors	• Play central role in strategic leadership and stewardship • Select, recruit, oversee, evaluate and compensate CEO/senior management • Be involved in board composition, diversity and nomination
Empowerment & accountability	• Empower board • As institutional investors, are accountable to individual savers (e.g., mutual and pension funds)	• Delegate sufficient authority to CEO (and committees) • Be accountable to owners at annual general meeting and special meetings
Communication & transparency	• Communicate expectations to Board and CEO • Receive corporate and auditor reports	• Have central role in ensuring comprehensive corporate communications plan • Speak with one voice • Evaluate and share information needs of Board
Service & fairness	• As investors, support efficient and transparent capital markets • Make social and ethical investments to promote corporate social responsibility	• Balance diverse expectations of shareholders and other stakeholders • Champion integrated social responsibility as part of overall mission of company
Accomplishment & measurement	• Allocate resources (especially capital) among competing corporations • Approve fundamental changes affecting corporation • Use corporate performance results	• Oversee corporate performance • Approve certain important decisions of management or executive • Use performance measures and verify their integrity
Learning & growth	• Reward learning cultures with investment • Contribute to growth through the workings of capital markets	• Foster climate of continuous learning and growth • Be committed to director development

Source: Adapted from Conference Board (1998), part of Exhibit 8, page 8.

who does what in the new corporation will not be the same as in the old. What is particularly important about changing governance models is the re-definition of roles to be played and activities to be conducted. Again, for a typical for-profit corporation, these are defined in Table 7.2.

In implementing port devolution policies, it is critical for governments not moving all the way along the continuum towards private ownership to carefully pre-determine how 'owner' functions will be handled in order to ensure that their approaches have integrity. In sum, the critical question for those implementing the not-for-profit model becomes: who fills the owner (shareholder) role in a non-share capital corporation?

There is little evidence that the issue of roles (and responsibilities) was as carefully considered by the Government of Canada when it devolved ports in Canada as when it devolved airports. Therefore, while there are no positive lessons to be learned from the Canadian port experience, there are lessons to be learned from the governance literature.

What are the overarching Codes of Governance that governments need to put in place to ensure the success of the devolution process? The TSE-14 provides one example. It has been adopted by non-profit organizations and could also be made mandatory for non-profits reporting to stakeholders. It may be asked: if companies are not publicly traded, are Codes of Governance needed at all?

Sachs et al. (2000) examine the relationship between privatization, institutional reforms and overall economic performance in transition economies, distinguishing between 'change-of-title' privatization and 'agency-related' reform. They argue that not only is the former insufficient to generate economic performance improvements without the latter, but also that privatization may even have negative economic performance impacts. Agency-related reforms include improving the performance and regulation of capital markets, and laying the foundation for appropriate corporate governance. They conclude that implementation of Codes of Good Governance is a necessary condition for maximizing of benefits of devolution.

The OECD (1999) has developed five key features of the regulatory climate for good governance in corporations:

- The corporate governance framework should protect shareholders' rights.
- The corporate governance framework should ensure the equitable treatment of all shareholders, including minority and foreign shareholders. All shareholders should have the opportunity to obtain effective redress for violation of their rights.
- The corporate governance framework should recognise the rights of stakeholders as established by law and encourage active co-operation between corporations and stakeholders in creating wealth, jobs, and the sustainability of financially sound enterprises.

- The corporate governance framework should ensure that timely and accurate disclosure is made on all material matters regarding the corporation, including the financial situation, performance, ownership, and governance of the company.
- The corporate governance framework should ensure the strategic guidance of the company, the effective monitoring of management by the board, and the board's accountability to the company and the shareholders.

Aguilera and Cuero-Cazurra (2000) maintain that increasing privatization has resulted in the development of Codes of Good Governance, such as the TSE-14 discussed above. They argue that these serve to compensate for deficiencies in the legal system that protects shareholder rights.

In conclusion, this section of the chapter has found that: (i) the implementation of a Code of Governance is a necessary step for a government to take to maximize the benefits of devolution; and (ii) if a pure private sector model is not to be followed, the government must be clear in its structuring of the devolved entity how the owner (shareholder) role is to be reallocated, and ensure that the reallocation of the role results in a congruent, internally consistent model for behaviour.

7.4 Implementing good governance principles within the organization

Why should good governance practices be adopted by corporations? According to the Conference Board (1998: 3):

> Companies that excel in governance practices post higher long-term profit growth, experience faster sales increases and are much more likely to be the leading companies in their sector.

This belief is further supported by The Dey Report (TSE, 1994: 2).

> We believe that effective corporate governance will, in the long term, improve corporate performance and benefit shareholders. Improved corporate performance is not only in the best interest of shareholders but also services the public interest generally.

While governance principles may provide guidance to the board of any transport entity, the outcome of good governance practices is dependent on the implementation structures and systems, and their effectiveness in achieving the desired outcome. This is easier said than done.

It is necessary that the board first come to some agreement about what principles are applicable to the organization, whether it is a public or private entity. Clearly, this must be undertaken in the context of the mission of the organization, its goals and objectives. Board members should be active participants in the process; this is not solely an activity for management divorced from the board.

Secondly, once consensus has been achieved, that agreement must be articulated and processes to facilitate effective decision making determined. The most common approach is to develop a reference document for board members. It should address a number of areas important to good governance practice, including appropriate board composition, the selection of directors, processes for such selection and appointment, appropriate board compensation, and terms of appointment. As part of good governance practice, it should include appropriate guidance for committee structure, size and activities as well as guidelines for committee reporting and decision making. Other necessary elements include processes for director removal/ replacement or reappointment, processes for stakeholder input and the resulting documentation should clearly delineate the responsibilities and obligations of directors contrasted with those of management.

Finally, governance decisions should be transparent and directors should be held accountable for outcomes. Without transparency and accountability, boards will fail to be responsible for their actions and hence the outcomes.

Given these generalizations about implementing good board governance practice, the question then arises: how can government ensure appropriate economic development outcomes once ports are devolved. What can be found in existing port devolution studies to answer this question?

7.5   The port devolution literature

As noted earlier, there is very little published research on port devolution, and most that has been written only examines port privatization. This is quite surprising given the worldwide trend towards new public management and the subsequent devolution of government-owned entities using a wide variety of devolution models, only one of which is privatization. Belief in the privatization model is encouraged by findings, like those of Boardman and Vining (1989) that, in terms of profitability, public sector firms perform substantially worse than private sector firms.

Gillen and Cooper (1995) provide an excellent summary of the rationale for port privatization. They point out that governments believe it encourages and improves efficiency, makes industry more responsive to the demands of customers, reduces public debt, and forces management (and unions) to face the realities of the marketplace. However, the record is not so rosy. According to Thomas (1994), this was the thinking behind UK port policy changes beginning in 1979 and culminating in the trust port privatization programme examined by Baird (1995) and Saundry and Turnbull (1997). These two studies conclude that the outcome of the programme was less than stellar. Saundry and Turnbull (1997) note that trust port privatization was accompanied by large profits made by former managers, and conclude that privatization did not transform the financial and economic performance of UK ports sufficiently to justify the private gains of port management shareholders. They attribute improvements in economic performance instead to the abolition of the National Dock Labour Scheme. Baird (1995) concludes that many of the trust ports were sold at below real market value and that, in some cases, there were no meaningful competing bids. On the other hand, Shashikumar (1998) reports efficiency improvements in the order of 90 per cent in the first six months after the privatization of the Indian port of Bombay (Mumbai), attributing its apparent success to the adherence of the parties involved to the principles of privatization.

What about non-privatization models and the appropriate role for port authorities? This has been open to discussion over the past decade (see, inter alia, Goss, 1990; Everett and Robinson, 1997; Grosdidier de Matons, 1997; Baird, 1999). Under the enterprise management system, Australian ports were unable to deliver aggressive port management because they failed to alter the fundamental structure of port authorities; the culture within port authorities discouraged innovation and initiative (Dick and Robinson, 1992). Ircha (1999) explores the derailing of the Canadian port devolution process by political interference. Baltazar and Brooks (2001) attribute the poor Canadian outcome to a misalignment of the strategy–structure–environment configuration. They conclude that what is necessary is internal consistency (called fit) between the environment, a government's goals, a port's strategy, and the structure and systems put in place at the time of devolution, a model they call the Matching Framework. Neither of the two non-privatization cases they present – Canada and the Philippines – is predicted to be successful because fit is missing.

As Juhel (2001: 174) notes, 'the new distribution of roles between public and private actors, in particular, calls for an appropriate allocation of duties and responsibilities, of risks and rewards, to make the global transportation system work to its best efficiency.' For Juhel, the balance between public and private sectors in industries contributing to economic development is the critical issue in performance outcomes. For Baltazar and Brooks (2001), good performance outcomes require fit. They contend that full privatization is not necessary and that alternate devolution models can be successful if fit is present.

7.6 Ensuring appropriate economic development outcomes

The key issue in the economic development equation is how to ensure that appropriate value-added activities take place within the boundaries of the local community rather than at another geographical location. This is really an issue of aligning the port's strategy and goals to those of the local community. It is possible to secure such alignment in part through the structuring of the organization by government in the devolution process; for more on appropriate organization structure for ports, see Baltazar and Brooks (2001). A fully satisfactory outcome lies in a combination of good planning, good timing and good luck. The best decisions are not guaranteed as the players are human and subject to all the potential frailties implied. There are, however, several issues to be addressed in ensuring the Board meets local community objectives (of economic development).

Development of an effective, independent board and avoidance of political patronage

As noted above, the role of the board is to set the vision and policy of the organization, while the role of management is to execute the strategic plan established by the board. What makes an effective board?

While there has been much discussion about implementation minutiae, such as term limits for directors and conflict of interest declarations, Leighton and Thain (1997) concluded that an effective board begins with an effective chairman. In the for-profit board, the board elects the chairman from among its members; therefore, the quality of the chairman can only be as good as the quality of the board from which the chairman is recruited. Crawford et al. (1993) argued that the most critical factor in board performance is the composition of the board; that is, if its members have integrity, experience and dedication, it will be effective. Leighton and Thain (1997) argued that boards must be empowered, responsible, effective and involved or the outcome will not meet the desired standard.

Thain and Leighton (1995) identified six factors behind board failure or success: legitimacy and power, job definition, board culture, competence, board management and board leadership. The last four can be influenced through board selection, and the choice government makes about whether the board will be politically chosen or drawn from the community, or a mix of these. Moreover, research by Grobmyer (2000) concluded that the skill mix of board members is a key to board effectiveness. The case of airport devolution in Canada illustrated the importance of allowing Boards the opportunity to backfill missing skills through the appointment of directors serving at the pleasure of the board (Brooks and Prentice, 2001).

One particular feature of devolution is that a government often establishes boards with a membership, structure and mix that it thinks is appropriate from a public policy perspective. It can only be concluded that governments, by the way they establish boards, set them up for success or failure. If a board appointment is seen as a plum or reward, as happens in cases of political patronage (government-appointed boards), effectiveness will be compromised. The board member who is more interested in fees than outcomes, in ego than results, and in political gain than community service or improving shareholder value can derail a community-driven board quite effectively. There is no place for political patronage if boards are to be effective. This does not mean that a government-appointed Board will not be effective; the key is the motivation and rationale for the choice of the particular board member. The process of selection is therefore also important.

Removal of the profit motive while retaining efficiency incentives

Privatized entities, finding themselves in a monopoly position, may not actually pass the benefits achieved to customers in the form of reduced charges or improved services, as noted by Saundry and Turnbull (1997) and Baird (1995) above. In the Gellman Research Associates (1990) study of the distribution of airport 'profits', it was concluded that US airports, because of their municipal ownership, pass their profits onto airlines in the form of below-market prices. The report recommends that governments limit excess profits through tax policy, rather than regulating the 'explicit profits' seen in for-profit airports.

In Canada, the choice of implementing devolution of airports via non-share capital corporations was considered by the government to be preferable. The not-for-profit non-share capital corporation model encourages corporations to follow one of three paths:

- allocating positive operating returns to capital projects, which may or may not be necessary,
- allocating positive operating returns to reduced services fees, or
- expending positive operating returns by enlarging the administrative bureaucracy (because it adds to local employment).

The path chosen is a function of the philosophy of board members and what they deem their objectives to be.

The second part of the balancing act is retaining or improving efficiency without the profit motive. In a private sector operation, efficiency gains are driven by the prospect of profit. The motivation for such gains in the public/private model or the non-profit model is unclear. If they are achieved, efficiency or productivity gains will contribute to positive returns, which then result in the options noted above. Unless articulated by government as part of the devolution policy (and its implementing legislation) or by the community as part of its vision, the outcome may not be the option that best supports economic development.

Alignment of board vision and objectives with community objectives

In the process of devolution, each devolved entity faces an identity crisis: does its board co-opt the objectives of government, or identify its own in keeping with the views of the directors of the newly created board, or those of its community's stakeholders? The outcome will very obviously depend on how directors are chosen. It is highly likely that those inserted by government – be it national or local – will co-opt the objectives of the appointing government if in the majority. Those chosen by the community will more likely reflect the concerns of local entities, while those chosen by stakeholders have yet again different objectives. Early experiences no doubt play a role in influencing whether a particular board will co-opt the objectives of those its members represent or whether members feel sufficiently independent to adopt objectives of the Board's own choosing.

The board is the steward of the port's assets; its directors must exercise due diligence, meet fiduciary obligations and understand community economic development objectives. A review of organizational design literature indicates that a good board needs an orientation to community objectives, a solid reporting system to its community, guidelines on transparency in tendering, collegiality within and between the board and management. Before managers can make decisions that take into account stakeholder relationships, they must know which interests stakeholders consider to be important. Brooks, Prentice and Flood (2000) believe that stakeholder consultation is a minimum feature of good governance in ports and airports.

Development of mechanisms for board accountability

Most devolution models require boards to produce annual reports as a mechanism of accountability. Is this really adequate? Given the poor track record of Canadian for-profit corporations found by TSE (1999) and noted in the Appendix, the answer must be no. In its devolution of airports, the Government of Canada, introduced a set of public accountability principles that were mandatory for airports to follow (Auditor General of Canada, 2000). These are:

- Not-for-profit corporation.
- Board of Directors includes two or three federal nominees.
- Equitable access to all carriers.
- Reasonable user charges.
- Engage in activities consistent with its purpose.
- General practice to tender contracts.
- Declarations of business activities to avoid real or perceived conflict of interest.
- Community consultations.

Of these, perhaps the most important is the requirement for community consultations (defined as twice-yearly meetings with a formal community consultation committee representing specific interests). This does not address the question. accountable to whom? Brooks, Prentice and Flood (2000: 135) conclude that:

> It is the responsibility of government to ensure that the quality of the agreement signed in the process of devolution ensures responsible management. Accountability is no longer just political; newly created entities must be responsible to their community stakeholders, their financiers, as well as those they directly service, be they tenants, suppliers or users.

If the issue is one of aligning board objectives with economic development interests, community consultation is only the beginning.

7.7 Conclusions

How well have port devolution initiatives worked? Have they been compatible with local community economic development initiatives? To answer these questions, further research is needed. Because of the sheer complexity of port operations, and the wide variety of devolution models in use by governments, empirical research needs to be both transnational and case-based, a methodology that can only succeed if it is adopted widely and consistently throughout the maritime economics community. Otherwise, it will just continue to be anecdotal.

The majority of the research reported in the academic press examines corporate governance as it relates to public or private entities from the American/Anglo-Saxon market-based perspective. This chapter has done so as well. It is important to remember that there are alternative models of governance, such as those explored in a collection of essays edited by Chew (1997); these include the Japanese relationship-based model, one that does not have a key tenet of transparency but rather active control within complicated networks. Because of the wide variety and complexity of governance models, governments are unlikely to agree on a globally harmonized approach to the governance of ports. Such harmonization is not a necessary prerequisite to good governance at the port level: 'Good corporate governance can contribute to the corporation's achievement of both its public policy and commercial objectives' (Department of Finance, 1996: 1).

This chapter has suggested some of the issues governments need to contemplate in their quest for both good governance practices and the alignment of board commercial objectives with community economic development objectives. Without the former, the latter is less likely to occur.

It is the author's contention that the practice of good corporate governance, whether within a non-share capital port or a private port, is not incompatible with the ability of the corporation to deliver positive economic development outcomes. It is the mechanisms of accountability and transparency, including a good stakeholder communication policy, coupled with the commitment of board directors to serving the public interest, that secure the success of a port economic development agenda under non-government owned and – controlled models.

While governance practices are under scrutiny in the private sector, there is no reason to limit the application of Codes of Good Governance to private sector ports. The author contends that the application of such a code to any devolved port is the path to ensuring a greater likelihood that public policy objectives will be met.

It is the means by which the owner role, or the shareholder right, can be replaced by the community's interest, if the community is treated in a manner similar to the owner under the code. In the private sector, shareholders profit from good governance while poor management imposes costs on employees, customers, suppliers and the local community. In other governance models, poor management also imposes costs on the local community. As noted by Roe (1994: ix), 'society wins if governance works'.

Notes

1 Stakeholder theory (as developed by Berman et al., 1999; Alkhafaji, 1989; and Louma and Goodstein, 1999) deals with questions of corporate governance, e.g., the legitimacy of who serves on boards and the appropriate roles of boards vis-à-vis stakeholders.

References

Aguilera, R.V. and Cuero-Cazurra, A. (2000) *Codes of Good Governance Worldwide*, CIBER Working Paper in International Business No. 00-105, Urbana-Champaign, IL: University of Illinois at Urbana-Champaign.

Alkhafaji, A.F. (1989) *A Stakeholder Approach to Corporate Governance: Managing in a Dynamic Environment*, New York: Quorum Books.

Auditor General of Canada (2000) *2000 Report of the Auditor General of Canada* (Chapter 10: Transport Canada–Airport Transfers: The National Airport System), Ottawa: Office of the Auditor General of Canada (www.oag-bvg.gc.ca).

Baird, A.J. (1995) Privatisation of Trust Ports in the United Kingdom: Review and Analysis of the First Sales, *Transport Policy*, 2(2), 135–43.

Baird, A.J. (1999) Privatisation Defined: Is It the Universal Panacea?, 27 June. www.worldbank.org/transport/ports/con_docs/baird.pdf

Baltazar, R. and Brooks, M.R. (2001) *The Governance of Port Devolution: A Tale of Two Countries*, World Conference on Transport Research, Seoul, Korea.

Berman, S.L., Wicks, A.C., Kotha, S. and Jones, T.M. (1999) Does Stakeholder Orientation Matter? The Relationship Between Stakeholder Management Models and Firm Financial Performance, *Academy of Management Journal*, 42(5), 488–506.

Boardman, A.E. and Vining, A.R. (1989) Ownership and Performance in Competitive Environments: A Comparison of the Performance of Private, Mixed and State Owned Enterprises, *Journal of Law and Economics*, 32(1), 3–33.

Brooks, M.R., Prentice, B. and Flood,T. (2000) Governance and Commercialization: Delivering the Vision, *Proceedings*, Canadian Transportation Research Forum, June, 1, 129–43.

Brooks, M.R. and Prentice, B. (2001) *Airport Devolution: The Canadian Experience*, World Conference on Transport Research, Seoul, Korea.

Cadbury Commission (1992) *Code of Best Practice: Report of the Committee on the Financial Aspects of Corporate Governance*, London: Gee and Co. Ltd.

Carver, J. (1990) *Boards that Make a Difference*, San Francisco, CA: Jossey-Bass Inc.

Chandler, A.D. Jr. (1962) *Strategy and Structure: Chapters in the History of the American Industrial Enterprise*, Cambridge: MIT Press.

Chew, D. (ed.) (1997) *Studies in International Corporate Finance and Governance Systems: A Comparison of the US, Japan and Europe*, Oxford: Oxford University Press.

Conference Board of Canada (1998) *Canadian Directorship Practices 1997: A Quantum Leap in Governance* (225-98 Report), Ottawa: The Conference Board of Canada.

Crawford, P., Dimma, W., Powis, A., MacDougall, H. and Taylor, C. (1993) Dialogue: Improving Board Effectiveness, *Business Quarterly*, 57(3), 11–18.

Department of Finance and Treasury Board of Canada (1996) *Corporate Governance in Crown Corporations and Other Public Enterprises*, Ottawa.

Dick, H. and Robinson, R. (1992) *Waterfront Reform: The Next Phase*, Presented to the National Agriculture and Resources Outlook Conference, Canberra February.

Everett, S. and Robinson, R. (1997) *Reforming Ports: Issues in the Privatisation Debate*, Working Paper 97–7, Sydney: Institute of Technology.

Galbraith, J.R. and Kazanjian, R.K. (1978) *Strategy Implementation: Structure, Systems and Process*, 2nd edn, St. Paul, MN: West Publishing Company.

Gellman Research Associates (1990) Analysis of Airport Cost Allocation and Pricing Options, *Work Element III Report, Part II*, Jenkintown, PA: Gellman Research Associates.

Gillen, D. and Cooper, D. (1995) Public versus Private Ownership and Operation of Airports and Seaports in Canada chapter 1, in F. Palda (ed.), *Essays in Canadian Surface Transport*, The Fraser Institute, Vancouver (www.fraserinstitute.ca/publications).

Goss, R.O. (1990) Economic Policies and Seaports – Part 3: Are Port Authorities Necessary?, *Maritime Policy and Management*, 17(4), 257–271.

Grobmyer, J.E. (2000) Motivating Change: Expanding the Operations Assessment to Enhance Board Performance, *Health Care Strategic Management*, 18(7), 16–17.

Grosdidier de Matons, J. (1997) Is Public Authority Still Necessary Following Privatisation?, *Proceedings of the Cargo Systems Port Financing Conference*, London, 26–27 June. www.worldbank.org/transport/ports/con_docs/dematons.pdf

Ircha, M. C. (1999) Port Reform: International Perspectives and the Canadian Model, *Canadian Public Administration*, 42(1), 108–32.

Juhel, M.H. (2001) Globalization, Privatisation and Restructuring of Ports, *International Journal of Maritime Economics*, 3, 139–74.

Kaufmann, D., Kraay, A., and Zoido-Lobatón, P. (2000) Governance Matters: From Measurement to Action, *Finance & Development*, 37(2), International Monetary Fund, http://www.imf.org/external/pubs/ft/fandd/2000/06/Kauf.htm

Kazanjian, J. (2000) Assessing Boards and Individual Directors, *Ivey Business Journal*, 64(5), 45–50.

Keasey, K., Thompson, S. and Wright, M. (1999) *Corporate Governance*, Cheltenham, UK: Edward Elgar.

Leighton, D.S.R. and Thain, D.H. (1997) *Making Boards Work: What Directors Must Do to Make Canadian Boards Effective*, Toronto: McGraw-Hill.

Louma, P. and Goodstein, J. (1999) Stakeholders and Corporate Boards: Institutional Influences on Board Composition and Structure, *Academy of Management Journal,* 42(5), 553–63.

OECD (1999) *Principles of Corporate Governance,* (SG/CG(99)5), Paris: Organisation of Economic Co-operation and Development.

Parr, J.B. (1981) The Distribution of Economic Opportunity in a Central Place System: Dynamic Aspects and Growth Poles, in A. Kuklinski (ed.), *Polarized Development and Regional Policies: Tribute to Jacques Boudeville,* The Hague: Mouton Publishers.

Roe, M.J. (1994) *Strong Managers, Weak Owners: The Political Roots of American Corporate Finance,* Princeton, NJ: Princeton University Press.

Sachs, J., Zinnes, C. and Eilat, Y. (2000) The Gains from Privatization in Transition Economies: Is 'Change of Ownership' Enough? (Paper 63 CAER II Project), Cambridge, MA: Harvard Institute for International Development.

Saundry, R. and Turnbull, P. (1997) Private Profit, Public Loss: The Financial and Economic Performance of UK Ports, *Maritime Policy and Management,* 24(4), 319–34.

Shashikumar, N. (1998) The Indian Port Privatization Model: A Critique, *Transportation Journal,* 37(3), 35–48.

Sternberg, E. (1998) *Corporate Governance: Accountability in the Marketplace,* London: The Institute of Economic Affairs.

Thain, D.H. and Leighton, D.S.R. (1995) Why Boards Fail, *Business Quarterly,* 59(3), 71.

Thomas, B.J. (1994) The Privatization of United Kingdom Seaports, *Maritime Policy and Management,* 21(2), 135–48.

Toronto Stock Exchange and Institute of Corporate Directors (TSE) (1999) *Report on Corporate Governance, 1999: Five Years to the Dey,* Toronto: Toronto Stock Exchange. Toronto Stock Exchange Committee on Corporate Governance (TSE) (1994) *Where Were the Directors? Guidelines for Improved Corporate Governance in Canada (The Dey Report),* Toronto: Toronto Stock Exchange.

World Bank (undated) *World Bank Port Reform Toolkit,* www.worldbank.org/transport/port/toolkit/

Table of Principles	The Good Governance Perspective	Adoption Rate
1. The board should assume responsibility for the stewardship of the corporation.	This includes strategic planning, risk management, succession planning, communications policy, and internal control and management information systems. A board needs to set policy and not become involved in micro-management.	Good (strategic planning, risk management, and controls; to poor (succession and communications)
2. The majority of directors should be unrelated.	Independence of management is a key element in the ability of the director to monitor results.	Good
3. The board should provide disclosure of measures to determine relatedness.	Full disclosure of a director's relationships and interests ensures that relatedness is transparent.	Good
4. A committee of independent directors should be responsible for the appointment and assessment of directors.	Usually called the governance committee, this committee ensures that board appointments are not unduly influenced by management and the director performance is evaluated.	Poor
5. There should be a process for assessing board effectiveness, as well as the effectiveness of its committees and individual directors.	Such processes should measure performance against benchmarks so that the board can engage in continuous improvement in its stewardship.	Poor
6. The board should provide orientation and education for new directors.	Without orientation and education, it is unlikely that the full potential contribution of directors can be realized.	Fair
7. The board should consider reducing its size to improve its effectiveness.	Overly large boards are considered to be less effective in decision making. Not noted here, but of considerable importance in the literature, is the issue of board skill set and mix, which should reflect the range of skills needed, and the diversity of interests.	Excellent
8. The board should review director compensation in light of risks and responsibilities.	Fees should be commensurate with director's duties, responsibilities and risks.	Good

Appendix: The TSE 14 Principles – *continued*

Table of Principles	The Good Governance Perspective	Adoption Rate
9. Committees of boards should be composed of unrelated non-management Directors.	The ability of a board to address management shortfalls in execution of board policy is hampered when committees are controlled by management.	Good
10. The board should appoint a committee responsible for governance issues.	Without a committee of the board taking responsibility for governance, governance practices will likely fail to be adequately reviewed by the corporation.	Fair
11. The board should define the limits of responsibilities by establishing mandates for the board and the CEO in line with objectives.	It is difficult to measure performance if objectives and mandates have not been determined. It is the board's responsibility to clarify these within, and with the CEO.	Fair
12. The board should establish structures and procedures for the board to operate independent of management.	In keeping with maintaining board independence of management, appropriate structures often include such elements as a chair who is an unrelated director and the ability of the board to hold meetings without management present.	Good (unrelated chair) to poor (independent meetings)
13. The board should establish an audit committee, with a defined mandate, and independent of management.	Adequate oversight of management dictates the existence of an audit committee with responsibility and authority independent of management.	Good
14. The board should implement a system whereby individual directors can engage independent advisors at the corporation's expense.	Such ability is seen as essential to the protection of independence. It is often suggested that a committee of unrelated directors should approve such expenditures.	Poor

Source: The principles are a paraphrase of TSE (1994) while the adoption rate is based on results from TSE (1999: 3) where the author has concluded that excellent = above 80%, good = above 60%, fair = above 40% and poor = below 40%. The governance perspective is a lay interpretation of the key issues.

8

Privatization Trends at the World's Top Hundred Container Ports

Alfred J. Baird

8.1 Introduction

The increasing intervention of the private sector in performing a number of essential activities related to the effective functioning of seaports has generally been an acknowledged trend in recent years (e.g. Thomas, 1994; Baird, 1995; Cass, 1996). However, in seeking to probe deeper into the phenomenon of what has become known as 'port privatization', certain questions arise that require further investigation (Goss, 1990; De Monie, 1994; Baird, 2000). Far from an exhaustive list, these questions might include:

- What is the extent of private sector intervention in seaports?
- Which specific seaport activities do the public and private sector perform?
- What methods of privatization are used, and what changes does this imply for the role of both the public and private sector?, and;
- What are perceived to be the main advantages and disadvantages of these changing institutional arrangements?

This chapter seeks to tackle these questions using the following approach. First, through a review of a survey designed to establish the extent of institutional reform in seaports, undertaken during 1998–99 by the International Association of Ports and Harbours (IAPH, 1999). This survey highlights the split of seaport activities and responsibilities vis-à-vis public and private sector organizations for most of the world's significant ports.

Secondly, a further survey of seaports is undertaken, this time by Napier University. This latter survey, focusing on the world's top one hundred container ports, identifies the objectives and methods used by ports to effect privatization, establishing the share of investments made by private and public entities over recent years. The survey also considers the perceived advantages and disadvantages, from the port point of view, of increased private sector intervention, identifying the role of the port authority in the midst of ongoing institutional changes.

Findings from the two surveys are considered to be complementary. The IAPH survey offered an essential basis and starting point for the subsequent Napier survey, and raised questions and issues requiring further exploration. The Napier survey permitted some of these additional aspects concerning port privatization to be investigated.

The findings contained in this paper are expected to be of interest to the ports and shipping industry, to investors in ports, to policy makers, and to the maritime research community.

8.2   Private sector intervention in seaports

IAPH survey

During 1999, an IAPH task force, called the Institutional Reform Working Group and headed by Malcolm Ravenscroft, formerly of Associated British Ports (ABP), undertook a major survey of IAPH members (IAPH, 1999). The survey was intended to establish the overall extent of public and private sector intervention in seaports.

The survey enjoyed a very good response rate, with 188 ports replying, equivalent to over 80 per cent of IAPH membership, and demonstrating a consistent regional balance. Most of the world's major ports responded, as well as many medium-sized ports.

Specific parts of the IAPH survey, which are reproduced here in a different format, essentially relate to the split or extent of public and private sector intervention in ports in respect of the following three contexts:

- Port organization.
- Port assets.
- Port operations.

It should be noted that some of the results do not round to 100 per cent. This is probably due to some incomplete responses being received, and is not regarded as significant. Findings from the study were as follows.

Port organization

Some 92 per cent of ports responding to the IAPH survey were public organizations. Of these, 71 per cent were either a public agency or corporation, and 21 per cent were departments of government (Figure 8.1). Only 7 per cent of ports were private companies, and of these, more than two-thirds have a government shareholding, ranging from 60 per cent to 100 per cent. The overwhelming majority of seaports therefore appear to be in some form of public ownership.

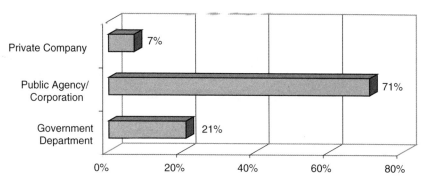

Figure 8.1 Port authority by organization type
Source: IAPH.

The few private ports where the state does not have a vested interest are primarily to be found in the United Kingdom, where the policy has aimed at achieving the outright disposal of port property rights, duties, and obligations, to private sector successor companies. Whilst the IAPH survey states that more ports are considering some form of privatization over the medium term (i.e. over the next five years), this is expected to relate to private sector provision of port assets and port services (see below), rather than to the outright transfer of port property rights as has occurred in the UK.

Nevertheless, when considering the survey findings, it is worth remembering that of the 'port authorities' mentioned, 7 per cent are private (or partially private), the remainder public.

Port assets

Figure 8.2 shows that, by and large, either the port authority or some other form of public organization owns port breakwaters and access channels. Private ownership of these particular assets on a worldwide basis appears to be, at best, negligible.

The reason for this may be largely due to the difficulty private companies would tend to experience in seeking to recover costs for breakwaters and channels, in addition to the public good nature of such assets (e.g. flood protection).

Analysis of port terminal and crane ownership in the IAPH survey is broken down further into the following three categories:

- Container terminals.
- Bulk terminals.
- General cargo terminals.

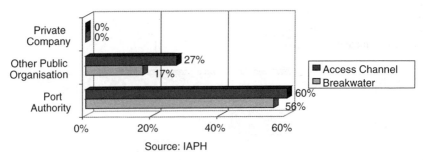

Figure 8.2   Ownership of port assets
Source: IAPH.

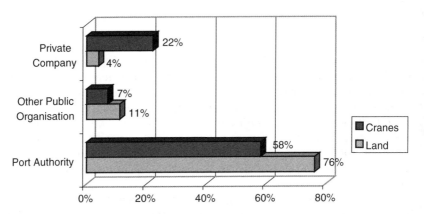

Figure 8.3   Ownership of port land/cranes used for container terminals
Source: IAPH.

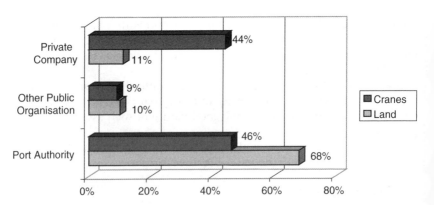

Figure 8.4   Ownership of port land/cranes used for bulk terminals
Source: IAPH.

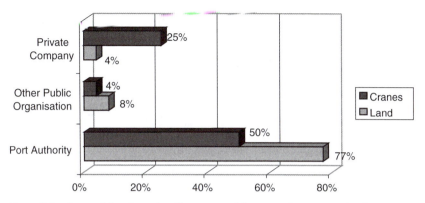

Figure 8.5 Ownership of port land/cranes used for general cargo terminals
Source: IAPH.

With regard to container terminals, almost 90 per cent of container terminal land within ports is either owned by the port authority or by another public body, with only 4 per cent owned by a private company (Figure 8.3).

Provision of container cranes is maintained by some 65 per cent of port authorities/public bodies, with container cranes owned by private companies identified at 22 per cent of ports.

In the case of bulk cargo terminals, port authorities own 68 per cent of port land, with some other public body owning a further 10 per cent, and private companies owning 11 per cent (Figure 8.4).

Provision of handling equipment at bulk terminals appears to be more evenly split between the port and private stevedores (55 per cent and 44 per cent respectively).

Port authorities also own the majority of cranes used in general cargo terminals (54 per cent), with the private sector owning 25 per cent of cranes (Figure 8.5). In terms of the ownership of land for general cargo terminals, 85 per cent of respondents stated that such facilities were owned either by the port authority or by another public body, with only 4 per cent owned by some other private company.

Port operations

For ease of analysis, the evaluation of public and private sector intervention in port services, based on IAPH data, has been split into three categories. These categories are:

- Port navigation services.
- Stevedoring services.
- Added value services.

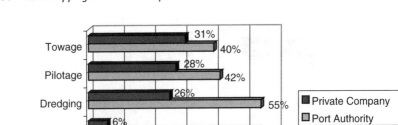

Figure 8.6   Provision of port navigation services
Source: IAPH.

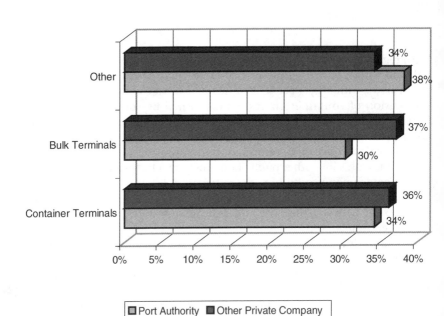

Figure 8.7   Provision of stevedoring services
Source: IAPH.

The breakdown of entities providing port navigation services is shown in Figure 8.6. The port authority (of which, to restate, 7 per cent are private, or partially private) provides navigation aids in 56 per cent of cases, the harbour master (54 per cent), dredging (55 per cent), pilotage (42 per cent), and towage (40 per cent).

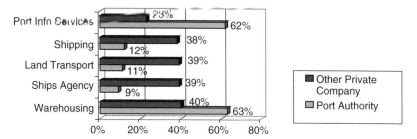

Figure 8.8   Provision of added value services
Source: IAPH.

Private companies provide navigation aids in 12 per cent of cases, harbour master (6 per cent), dredging (26 per cent), pilotage (28 per cent), and towage (31 per cent).

Unsurprisingly, private sector companies perform rather better with regard to the share of stevedoring services (Figure 8.7), although the role of port authorities is still significant. Of the ports responding, private stevedores run 36 per cent of container terminals, with 34 per cent operated by the port authority.

For bulk terminals, private stevedores operate 37 per cent, and port authorities 30 per cent. For 'other' terminals, port authorities operate 38 per cent, and private stevedores 34 per cent.

In regard to port added value services, port authorities have a significant role in providing warehousing and port information services, whilst private companies mostly provide other services such as ships agency, land transport, and shipping (shown in Figure 8.8).

8.3   Methods and impacts of private sector intervention in seaports

Napier University survey

The Napier University survey was intended to build on the extensive findings of the IAPH survey, albeit placing a more specific emphasis on private sector intervention in container terminals. The aim was to probe a little deeper, in order to:

• Consider the aims of privatization;
• Establish the methods of privatization used, and;
• Assess the split of port investment for the public and private sector.

Other aspects considered in the survey included assessing the importance of port labour reform in attracting private investment, identifying some of

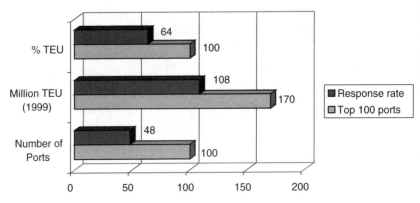

Figure 8.9   Napier port survey response rate
Source: IAPH.

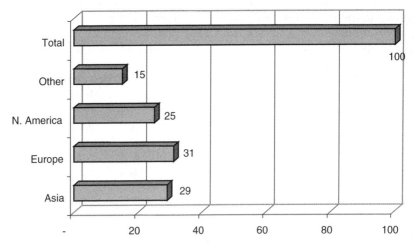

Figure 8.10   Ports responding by geographic region (percentages)
Source: IAPH.

the perceived advantages and disadvantages of private sector intervention from the port authority perspective, and establishing what the role of the 'new' port authority itself should be.

Each of the top hundred container ports received a questionnaire and covering letter, copies of which are shown as Appendix I at at the end of this chapter. As the top hundred container ports collectively account for an estimated 80 per cent of world container trade, it was thought that an extensive survey of these ports would provide sufficient data upon which generalizations concerning the remainder of the ports population might be made. Inevitably, however, this would depend upon the level of response.

Of the top 100 ports, a total of 48 ports returned the questionnaire (Figure 8.9). This level of response is regarded as reasonably satisfactory for such a study, and the findings are believed to offer a good indication of the overall picture in the container ports sector. Moreover, those ports that did respond collectively accounted for 108 million TEU in 1999, equivalent to 64 per cent of the overall throughput of the world's top one hundred ports, and equal to approximately half of world container port traffic.

Geographically, the survey responses yielded a balanced return with ports being representative of all the main economic regions (Figure 8.10). Of the 48 ports responding, 15 (31 per cent) were European, 14 (29 per cent) Asian, 12 (25 per cent) North American, and seven (15 per cent) from other regions.

The results of the survey are described in the following sub-sections.

Aims of privatization

By far the most common aim or motivation behind a port seeking to bring in the private sector is to increase efficiency, and consequently to lower port costs, with half of ports (50 per cent) mentioning this factor (Figure 8.11). Expanding trade as a specific aim of privatization was mentioned by 27 per cent of ports, and reducing the cost of investment to the public sector by 23 per cent. To obtain management know-how was mentioned by 15 per cent of ports.

A significant 21 per cent of ports mentioned a number of 'other' reasons or aims behind bringing in the private sector to operate and/or develop

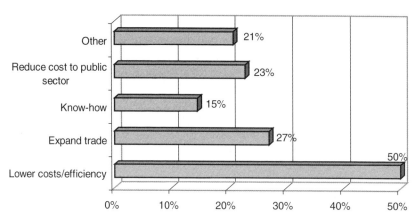

Figure 8.11   Aims behind bringing in private sector
Source: IAPH.

container terminals. These reasons included the speed of developing new terminals, complying with ports and harbour legislation, developing a public–private partnership, and increasing port revenue.

Methods of privatization

Terminal concessions and leasehold arrangements are the most common methods used by ports to facilitate private sector intervention, being used by 52 per cent of ports who responded to our survey (Figure 8.12). Build–operate–transfer (BOT) has been used by 19 per cent of ports, joint venture by 10 per cent, and outright sale of port land by 4 per cent.

Corporatization of a port authority (13 per cent) is generally combined with some form of concession/lease arrangement for terminal operations, with a number of ports mentioning both methods.

'Other' methods of privatization largely relate to shorter-term terminal rentals, or formation of a separate container terminal corporation, which might be wholly or partially owned either by the port authority or by some other public body.

Public and private sector investment

The survey sought to establish the approximate public/private investment split in container terminals over the past five years (1995–2000).

The graph in Figure 8.13 shows the split of private/public investment covering a range of values from under US$25 million to over US$250 million. Overall, and across all values, the results indicate that both public and private sector organizations are significant investors in port container terminals.

In approximately half of all ports, the level of investment over the last five years exceeds US$100 million, for both public and private sector entities, with absolute investments (for both sectors) appearing to be relatively equal.

What the data also indicate is that the public sector matches – and in some cases exceeds – private sector investments made in container terminals. Moreover, this excludes the cost of creating basic infrastructure (e.g. dredging, breakwater, etc.) which to a great extent is an additional cost met by the public sector.

This is not to say that all ports will necessarily have a mix of both public and private sector investment. On the contrary, some ports were found to depend largely upon private sector investment, whilst others were the opposite, relying much more heavily on public sector finance.

Labour reform

Respondents were asked to consider the importance of dockworker labour reform in attracting private sector investment.

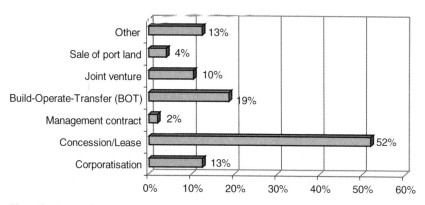

Figure 8.12 Methods of privatization employed
Source: IAPH.

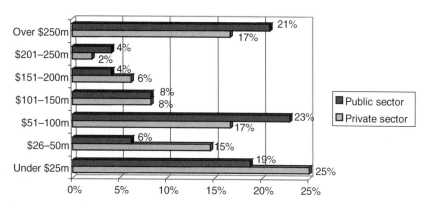

Figure 8.13 Private/public sector investment split in container terminals, 1995–2000
Source: IAPH.

Some 17 per cent of ports regarded labour reform as a critical element, with a further 30 per cent claiming it was either very important, or important (Figure 8.14). Overall, a slight majority of ports considered labour reform to be a significant issue. However, about 40 per cent of ports viewed labour reform to be of rather less significance, mostly comprising those ports at which labour had already been reformed (e.g. the United Kingdom, Italy, etc.).

There appeared to be no geographical similarity in regard to this aspect of the study. For example, while some ports in North America and in Asia regarded labour reform as critical, other ports in the same regions viewed it as being rather less important. It is not the aim of this study to

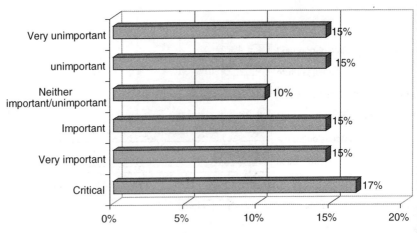

Figure 8.14 Importance of port labour reform for port privatization
Source: IAPH.

consider the reasons for such divergent views, but there is perhaps a need for further research to establish why there are different views within the same regions, albeit in different countries/states.

Advantages and disadvantages of privatization

The sharing of investment was considered by most ports (i.e. 50 per cent) to be the main advantage of private sector intervention, followed by benefits gained through improved productivity (44 per cent). Helping trade growth was mentioned by 38 per cent of ports, with management expertise mentioned by 31 per cent (Figure 8.15).

'Other advantages' of private sector intervention given by ports included making terminals profitable, keeping carriers in a port (e.g. where a carrier leases a terminal), competition between terminals (in a port), improved management and better facilitation of development.

The main disadvantages of private sector investment, according to ports, are shown in Figure 8.16. Most ports (31 per cent) stated the loss of control as an issue, with 21 per cent mentioning political and commercial ambiguity as a problem. Difficulties in operator selection (15 per cent), and the lengthy process for securing concessions etc. (8 per cent) were also highlighted.

Under 'other disadvantages', several ports stated that they did not perceive any disadvantages with private sector intervention. However, some ports mentioned other disadvantages such as inadequate income for the state, the possibility of an oligarchy developing, difficulties coordinating public and private investments, and the potential for unfair competition or preferential treatment.

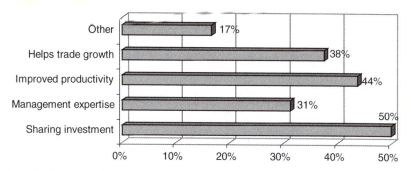

Figure 8.15 Main advantages of private sector investment in port
Source: IAPH.

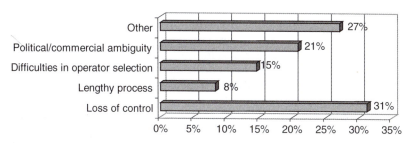

Figure 8.16 Main disadvantages of private sector investment in ports
Source: IAPH.

Role of the public sector port authority

A significant 63 per cent of ports believed that the role of the public port authority should include creating basic infrastructure (Figure 8.17). Other roles for a public port authority included overseeing port regulation and safety (46 per cent), ensuring fair competition and pricing (42 per cent), and protecting the public good aspect (40 per cent).

The public good aspect of a port might be expected to include flood protection measures, the presence of 'free-riders' (e.g. leisure and small craft freely using navigation channels), and protecting the strategic importance of the port to the economy and for national security.

Additional functions that a public port authority was expected to undertake included planning and marketing (25 per cent), and monitoring efficiency (19 per cent).

Under 'other' roles, where appropriate a public port authority was expected to reflect a regional, as opposed to a local, viewpoint, to finance terminals and cranes in certain instances, to enhance trade facilitation, and provide property management.

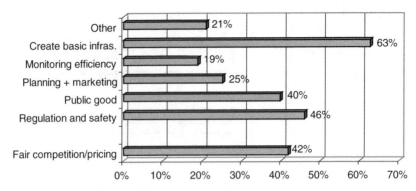

Figure 8.17   Role of the public sector port authority
Source: IAPH.

Conversely, the few wholly privately owned ports in the sample tended to criticize the role – and indeed the very existence – of a public port authority. The main reasons given to justify such a view included the lack of commercial focus, bureaucratic inefficiency, delay, and uncertainty, and a 'civil service' attitude, which restricted entrepreneurial development. Curiously, the port stating the latter impediment also argued that the role of a public port authority should include creating basic infrastructure.

8.4   Conclusions

The IAPH survey highlighted a number of issues with regard to changing institutional arrangements in the context of seaports. The main findings of the study are summarized as follows:

• The port authority is to all intents and purposes generally a public body, with very few exceptions.
• The port authority, or some other form of public body, will generally hold title to virtually all port land/terminals and access channels, with very few exceptions.
• Private companies are increasing their role in providing cargo-handling equipment, but the role of the port authority/public bodies in this regard is still significant.
• The private sector now has a major role in the provision of stevedoring services, although the port authority/public bodies also play a significant role in these operations.
• Provision and maintenance of navigation aids, channels (dredging) and harbour master responsibilities are still predominantly port authority/ public body activities. However, private companies appear to be increasing their role, particularly in pilotage and towage.

- The port authority/public body is largely responsible for provision of warehousing and port information services, whilst the private sector provides most other port added value services.

The Napier survey sought to build and extend upon the IAPH survey, albeit more specifically in the context of the top one hundred container ports. The objectives of the Napier study were to consider the aims of privatization, the methods of privatization, and the investment role of both private and public sector organizations over the past five years. The Napier survey also sought to probe some of the advantages and disadvantages of increasing private sector intervention in seaports. The principal findings from this survey were as follows:

- Lower port costs and improved efficiency (stated by 50 per cent of ports), expanding trade, and reducing the dependence on public sector investment, appear to be the main aims of port privatization.
- Terminal concession or lease (52 per cent of ports) is the most common method of privatization, followed by BOT, joint venture, and terminal rental.
- Investments in container terminals amongst the top hundred ports appear to be relatively evenly split between private and public sector. However, this tends to mask the reality in that, while some ports do have a mix of public and private investment, others depend mainly on private sector capital, whereas other ports depend mainly on public sector investment.
- In attracting private sector investment, almost half of ports (47 per cent) consider port labour reform to be either critical, very important, or important, whereas ports at which some reforms have already occurred view this issue to be rather less important.
- The main advantages to the port of private sector investment are sharing costs (50 per cent), improved productivity (44 per cent), and helping trade growth (38 per cent).
- The main disadvantages of private sector investment are loss of control (31 per cent), political and commercial ambiguity, difficulties associated with selecting an operator, and the lengthy process this entails.
- The role of a public port authority is considered to include creating basic infrastructure (63 per cent), regulation and safety (46 per cent), ensuring fair competition and pricing (42 per cent), and the public good (40 per cent).

A key conclusion from this research is that there does not appear to be a single, common, standard approach to port investment/port privatization. Yes, some approaches or models are used more than others, and particularly the public port authority/private concession or lease arrangement.

But, in general, the method used, and the public/private investment split, will depend on a range of factors. Such factors will inevitably include prevailing local/national laws, the local 'way of doing things', the level of demand/supply, and the extent and nature of competition.

However, while the evidence suggests significant involvement of the private sector, especially in port operations and services, this does not detract from the fact that the public sector, in virtually all instances, takes much more than just a passing interest in its seaport system. (For additional reasons supporting continued public sector involvement in seaports see IAPH, 1999; Grosdidier de Matons, 1997; Goss and Stevens, 2001; Hoffmann, 2001.) Whether through a port authority, marine department, or other body, in the vast majority of countries the public sector retains a central role in seaport planning, regulation, development and investment.

References

Baird, A. J. (1995) Privatization of Trust Ports in the United Kingdom: Review and Analysis of the First Sales, *Journal of Transport Policy*, 2(2), 135–43.

Baird, A. J. (2000) Port Privatization: Objectives, Extent, Process, and the UK Experience, *International Journal of Maritime Economics*, 2(3), 177–94.

Cass, S. (1996) *Port Privatization – Process, Players and Progress*, London: IIR Publications/Cargo Systems.

De Monie, G (1994) Mission and Role of Port Authorities, *Proceedings of the World Port Privatization Conference*, London, 27–8 September.

Goss, R. O. (1990) Economic Policies and Seaports – Part 3: Are Port Authorities Necessary?, *Maritime Policy and Management*, 17(4), 257–71.

Goss, R. and Stevens, H. (2001) Marginal Cost Pricing in Seaports, *International Journal of Maritime Economics*, 3(2), 128–38.

Grosdidier de Matons, J. (1997) Is a Public Authority still Necessary Following Privatization?, *Proceedings of the Cargo Systems Port Financing Conference*, London, 26–7 June.

Hoffmann, J. (2001) Latin American Ports: Results and Determinants of Private Sector Participation, *International Journal of Maritime Economics*, 3(2), 221–41.

IAPH (1999) Final Report of the 2000 Task Force Institutional Reform Working Group, *Proceedings of the International Association Port & Harbours* (IAPH) *World Ports Conference*, Kuala Lumpur, May.

Thomas, B.J. (1994) Privatization of UK Seaports, *Maritime Policy and Management*, 21(2), 135–48.

Appendix I (Letter and Questionnaire to Ports) For the Attention of the Port General Manager, Managing Director or Chief Executive

Survey Analyzing Trends in (Container) Port Privatization

Napier University is undertaking a brief survey of the top 100 container ports to further assess recent trends regarding port privatization. Findings from the survey will be presented at the 5th Annual PDI Financing and Investing in Ports Conference in London, between 18–19th September 2000.

We would respectfully request your valued assistance in this study by completing the short attached 1-page questionnaire and faxing it back to us. The questionnaire is very brief and is expected to take only a few minutes to complete. Port anonymity will be respected as all responses with be aggregated together to present overall findings.

All ports returning a completed questionnaire will receive in due course a full copy of the survey results from Napier University.

We do hope you will be able to assist us in this study and look forward to receiving your completed questionnaire in due course.

Yours truly,

Alfred J. Baird

Head, TRI Maritime Transport Research Group

ANALYSING TRENDS IN (CONTAINER) PORT PRIVATISATION
(QUESTIONNAIRE)
PLEASE FAX BACK TO A. BAIRD, NAPIER UNIVERSITY
FAX–OO 44 131 455 3484 (<u>OR</u> 3486)
Name of Port
(Please circle your preferred answers)

1. What were the main aims behind bringing the private sector into your container terminal operations?

Lower costs/ efficiency	Expand trade	Know-how	Reduce cost to public sector	Other

2. What public sector method(s) of privatisation have been used to develop your container terminals?

Corpora-tisation	Concession	Management contract	Build Operate Transfer (BOT)	Joint-venture	Sale of port land	Other

3. What is the approximate total value of **private sector** investment in your container terminals during the past 5 years?

Under $25m	$26–50m	$51–100m	$101–150m	$151–200m	$201–250m	Over $250m

4. What is the approximate total value of **public sector** investment in your container terminals during the past 5 years?

Under $25m	$26–50m	$51–100m	$101–150m	$151–200m	$201–250m	Over $250m

5. How important has been port labour reform in attracting private sector investment for your container terminals?

Critical	Very important	Important	Neither important nor unimportant	Unimportant	Very unimportant

6. What are the main **advantages** of private sector investment in container terminals?

Sharing investment	Management expertise	Improved productivity	Helps trade growth	Other

7. What are the main **disadvantages** of private sector investment in container terminals?

Loss of control	Lengthy process	Difficulties in operator selection	Political and commercial ambiguity	Other

8. Do you believe that there is a role for a public sector port authority, and if so, why/why not?

YES (answer below)		NO (Please give reason)				
Ensure fair competition/ pricing	Regulation and safety aspect	Public good marketing	Planning and	Monitoring efficiency infrastructure	Create basic	Other

9.Any other comments

9

A New Paradigm for Hub Ports in the Logistics Era

Dong-Wook Song and Tae-Woo Lee

9.1 Introduction

The rapid increase in world trade in the last decade has triggered a new round of port development, and caused the restructuring of the world port network, as well as more intensive inter-port competition (Robinson, 1998). In this process, the term 'hub port' has been frequently used, as emerging large container ports, such as the Port of Tanjung Pelepas (PTP) and Port Klang in Southeast Asia and Shenzhen in South China, proclaim their status in the region. As a consequence, carriers have more choice in the location of their container transhipment centres. The decision of Maersk Sealand and Evergreen to leave Singapore for PTP is one of the most recent examples of such a trend in the international maritime industry (*Lloyd's List Maritime Asia*, 2001).

In addition, there are also dramatic changes in the mode of world trade and freight transportation, characterized by the prevalence of business-to-business (B2B) transactions and integrated supply chains. In the port industry, these changes have been embodied in the increased demand for value-added logistics services and the integration of different transportation modes (Robinson, 2001). Thus, logistics services in a port are a contentious issue in port policy and management, as mega ports such as Hong Kong and Singapore regard logistics services as a key area to support their long-term vision.

With the above background in mind, this chapter attempts to clarify the dimensions and/or factors being considered as quality for hub ports in the logistics era, and to conceptually establish a new framework on which ocean carriers and port operators may develop their own reasonable criteria for the matter. This piece of work will be of interest to parties engaged in the port industry as the industry is facing an ever-changing business environment.

9.2 The concepts of 'hub port' and 'load centre'

The term 'hub' is usually used to describe the centre of a hub-and-spoke structure, which is commonly found in transportation, telecommunications and other network industries. A hub-and-spoke network is characterized by the concentration of cargo or information flow on the inter-hub link, and is generally attractive to transportation companies because of its widely perceived benefits of scale economies and consequent cost savings derived from concentrating flow density on network linkages between hub locations (Hormer and O'Kelly, 2001). In the shipping industry, the advantages of the hub-and-spoke system have been widely recognized. O'Mahony (1998) states that hub-and-spoke strategies have enabled the major carriers to achieve wider port coverage in geographical terms and to deploy larger ships on the major trading routes.

However, the precise meaning of the term 'hub' depends on the industry concerned. For example, in air transport, the US Federal Aviation Administration defines a large hub as an area which enplanes at least 4,283,192 passengers (Taaffe et al., 1996). In the maritime sector, on the contrary, a hub port is a loose and diverse concept, which is sometimes referred to as a mega port, a direct-call port, a hub port and a load centre port, a mega hub (greater than four million TEUs per annum), a super hub (greater than one million TEUs per annum), a pivot port, and so on (Alderton, 1999). Scotti (1998) considers 'a hub terminal' as a strategic place, which can potentially provide shipping lines with a transhipment centre between major markets. Robinson (1998) points out that hub locations have emerged as the articulation points between networks based on feeder shipping and networks based on mainline services or as the articulation points between mainline shipping networks and transcontinental rail nets. Based on the above discussion, this chapter defines a hub port as an area serving such functions as a transhipment centre and a gateway for the larger hinterlands by connecting mainline services with various feeder networks.

Of the concepts that are relevant to a hub port, the term 'load centre' should be also stressed as it has been theoretically developed to describe the phenomenon of container traffic concentration in a few, larger ports (Hayuth, 1978) and has been used to refer to ports where container traffic is consolidated. Moreover, as Alderton (1999) mentions, a hub port is sometimes treated as a load centre. However, strictly speaking, there is a difference between the two concepts. Cullinane (2000) argues that from a carrier's perspective, the adoption of the load centre concept means focusing their operations in certain ports, and that the advantages of the load centre concept appear at the carrier level, rather than at the port level. This argument implies that the load centre is a concept that is more appropriately associated with a carrier's routing decision than the description of a port's status. Nevertheless, since a hub port is a strategic location for the concentration of container traffic, it is an ideal place for load centre and container operation. Thus, the load centre has important characteristics of a hub port in terms of cargo flow; the quality of a load centre should be considered as an indispensable part of the quality of a hub port.

9.3   Literature review on hub port quality

To establish a new framework or paradigm of the quality of a hub port, we should identify the dimensions and factors that are regarded as being the components of quality. In previous studies, there are three main kinds of research that help us to develop a picture of them:

- The first category is studies of port selection criteria, in which a wide range of dimensions and factors have been listed and examined. The findings of this research, to a large degree, reflects the quality of a hub port.
- The second type is studies of the load centre concept, which is, as was mentioned in the previous section, the fundamental feature of a hub port and its quality should be regarded as an aspect of quality for a hub port.
- Finally, the third is studies on 'existing' hub ports. This kind of research reveals the factors attributed with the success of the hub ports, which can again be generalized and taken into account when establishing a new framework for the quality of a hub port.

In addition to the above three types, there are also other studies focusing on a single aspect of hub characteristics. One such example is the study by Fleming and Hayuth (1994) on the spatial characteristics of transportation hubs, which indicates that 'centrality and intermediacy' are the most important spatial qualities of both air and sea hubs. In the rest of this section, a review is conducted on the dimensions and factors reflecting the quality of a hub port.

Port selection criteria

A wide range of methodologies have been used in previous research on port selection criteria. In the late 1970s, Foster (1978) conducted two surveys to identify the factors that were of frequent concern to shippers when they were selecting a port. In his research, the factors were divided into three dimensions – that is, general factors, port services and ocean carrier services. All of the factors were examined and ranked according to scales derived from the shippers. It was found that 'costs of transport and port charges', 'container cranes' and 'scheduled conference liner services' were the most important factors on the shippers' side. Later, Slack (1985) undertook a survey via interviews with forwarders and exporters to understand what factors were mostly considered in selecting a seaport in the container age, and two dimensions – general factors and port service factors – were examined. Since the late 1980s, a series of surveys and research have been conducted on port selection (Murphy et al., 1988; Murphy et al., 1989; Murphy et al., 1991; Murphy et al., 1992) with respondents ranging from shippers, ports, and carriers to international freight forwarders. Two dimensions –

Table 9.1 A summary of port selection criteria studies

Study	Methodology	Dimensions	Top three factors
Foster (1979)	Survey	General factors	Cost of transport & port charges
			Proximity to plant
			Number of sailings
		Port services	Container cranes
			Consolidation services
			Warehousing
		Ocean carrier services	Scheduled conference liners service
			Fully cellular container service
			Scheduled non-conference service
Slack (1985)	Survey	General factors	Number of sailings
			Inland freight rates
			Proximity of port
		Port service factors	Road and rail services
			Container facilities
			Tracing systems
Murphy et al. (1988) Murphy et al. (1992)	Survey	Port service factors	Has equipment available
			Low frequency of loss and damage
			Convenient pickup and delivery
Malchow and Kanafani (2001)	Disaggregate Analysis	Port location	Inland distance
			Oceanic distance

customer service issues and freight handling capabilities – were thoroughly examined in this series of studies and two factors were identified to be the most important ones through synthesizing the evaluation – that is, equipment availability and a low frequency of loss and damage.

In the latest attempt of its kind, Malchow and Kanafani (2001) developed a new method for determining port selection factors. A data analysis technique was employed to analyse the flows of four kinds of commodities in eight major US ports. Four factors – ocean distance, inland distance, sailing frequency and vessel capacity – were analysed and the results revealed that only two of these had a significant impact. The two imply that an increase in either the ocean distance or the inland distance between the port and the shipment's origin or destination serve to make a port less attractive, with a high degree of elasticity.

Table 9.1 lists various dimensions and their respective top three factors that were examined and identified in previous port selection research. It can be seen that the following six dimensions are widely recognized:

- Port infrastructure and port equipment.
- Port charges.
- Port service quality and scope.
- Port location.
- The carriers' service in port, and
- Inland transportation.

Load centres

Hayuth (1978) was the first to develop the load centre concept in an attempt to describe the phenomenon of the concentration of container traffic in a few, large ports. In his study, he divided the evolution of the container port system into five phases:

- Preconditions for change.
- Initial container port development.
- Diffusion, consolidation and port concentration.
- The load centre; and
- The challenge of the periphery.

He concluded that a load centre port should have qualities such as a large-scale local market, high accessibility to inland markets, advantageous site and location, early adoption of the new system, and an aggressive port management.

Later, Marti (1988) applied the load centre concept in the analysis of the evolution of container ports in the Pacific Basin and went one step further in the study of the quality of a load centre from the aspect of the port–carrier relationship. His study states that one of the most important factors, which relates directly to port industry modification, is the introduction of door-to-door or point-to-point rates by shifting the choice of port from a shipper to a shipping line, and thus ports no longer have exclusive control over cargoes generated in their immediate manufacturing area. Furthermore, a new methodology was introduced in his study. That is a 'shift-share analysis', which provides a means to recognize the evolutionary pattern of a load centre on a quantitative basis. Shift-share analysis is a purely qualitative technique which describes the trend of regional traffic concentration by calculating the change in the share of individual ports in the total throughput of a region during a certain period.

Later, the load centre concept has been applied in various regional container port system studies (e.g. Starr, 1994; Notteboom, 1997; Notteboom and Winkelmans, 1998; Wang, 1998) as well as in domestic container transportation (e.g. Slack, 1994). The theory on load centres has accordingly been improved and modified. Hayuth (1995), in a continuation of his former research, states that the most common characteristics of a load centre port are related to its location (i.e. good foreland and hinterland accessibility and large hinterland) operation (i.e. high productivity, frequent port of call, reasonable transportation and port-user costs, high cargo-generating effect and high level of intermodality) infrastructure (i.e. state-of-the-art infrastructure and superstructure, large back-up-space on terminal) and degree of integration (i.e. EDI). Later, Notteboom (1997) uses shift-share analysis and suggests more precise qualities for load centres such as:

- Regular port of call for RTW (round-the-world) liner services.
- Large volume of container traffic.
- High transhipment figures (feeder ships); and
- Substantial positive shift-effects in more than two of the periods observed.

The methodologies and identified characteristics of load centre studies are listed in Table 9.2 which shows similar dimensions and factors seen in the

Table 9.2 A summary of load centre studies

Study	Ports or region	Methodology	Identified characters
Hayuth (1978)	Ports in the United States		Large-scale local market. High accessibility to inland markets. Advantageous site and location. Aggressiveness of port management. Early adoption of the new system.
Marti (1988)	Pacific Basin	Shift-share analysis	High gain in differential shift.
Hayuth (1995)			Good hinterland accessibility. High productivity and frequent port of call. High level of intermodality. State-of-the-art port equipment. Reasonable costs. High degree of integration.
Notteboom (1997)	European Container Port System	Hirshmann–Herfindahl Index Gini coefficient and Lorenz concentration curve Shift-share analysis	Regular port of call of RTW. Large container traffic. High transhipment figures. Substantially positive shift-effects.

studies on port selection criteria. The difference, however, exists in that studies of load centres place greater emphasis on the volume of container traffic and also the share of the regional market of an individual port.

'Existing' hub port

In addition to the aforementioned studies, there are some studies about the quality of a hub port which are based on the successful experience of existing hub ports. Cullinane (2000) reviews the competitive position of the port of Hong Kong, from macro and micro points of view. In his study, frequency of sailings, freight rates and customs procedures are considered to be the main determinants of port choice in the South China context. Tongzon (2001) states that the key factors attributable to the success of the port of Singapore as a transhipment hub are:

- Strategic location.
- High level of port efficiency.
- High port connectivity.
- Adequate infrastructure.
- Adequate information-structure and
- A wide range of port services.

Gem (1998) concludes that the competitive edge of the port of Rotterdam is based on the following factors:

- Centrally located in Europe.
- Capacity to handle the largest vessels.
- A hinterland of 350 million consumers.
- Different container terminals (multi-user as well as dedicated).
- Excellent hinterland connections and multimodal transport facilities.
- Specialized distribution companies and cooperative custom facilities.
- Logistics centre in the port of Rotterdam (a so-called 'Distripark').

The success story of world-class hub ports has, to a large extent, reflected the quality of hub ports, but there are some individual factors that may be overemphasized, and all of the quality factors should be reassessed in today's changed environment.

From the above review, it can be summarized that the following dimensions and respective various factors have widely been regarded as the most important in port selection and load centre identification and have been testified to by the success of the established hub ports:

- Port infrastructure and superstructure.
- Hinterland accessibility.
- Port location.

- Port services quality.
- Port charges and total costs.
- Port regional market share; and
- Carriers' service in a port.

However, the deficiencies of previous research are obvious. The research on port selection criteria is biased by the interests of the authors, while the results of load centre research are inadequate in the dimensions examined. What is more, the quality of a hub port is constantly changing in response to changes in the environment. From Hayuth's (1978) study in the late 1970s to the studies of Cullinane (2000) and Tongzon (2001) there has been a significant change in the dimensions and factors of research. Today, as logistics prevails in the maritime sector and the port community has been restructured due to its integration in the logistics chain, the former dimensions and factors should be reassessed and new dimensions should be identified as determinants of the quality of hub ports in the new environment.

9.4   Development of transport logistics and hub port quality

The Council of Logistics Management (1998) defines logistics as:

> The process of planning, implementing, and controlling the efficient, effective flow and storage of goods, services and related information from the point of origin to the point of consumption for the purpose of conforming to customer requirements.

From this definition we can see that logistics has a broad meaning, which comprises both shipping and the port industry. A seaport has been regarded as a function in a logistics system and hub ports as a place of convergence of intermodal transportation (IAPH, 1996) and transhipment traffic. As such, a hub port has gradually turned into a logistics platform (Alderton, 1995). Consequently, the development of transport logistics will have a substantial influence on the quality of a hub port. This section is dedicated to the development of logistics and its embodiment in the transport sector, and then provides an insight into its impact on hub ports.

Outsourcing logistics services

The outsourcing of logistics functions to partners, known as 'third-party logistics providers (3PLs)', has become an increasingly powerful alternative to the traditional, vertically integrated firm (Rabinovich et al., 1999). Logistics outsourcing has been regarded as a rapidly expanding source of competitive advantage and logistics cost savings to many industries as it enables the firms to concentrate on their core competencies and gain

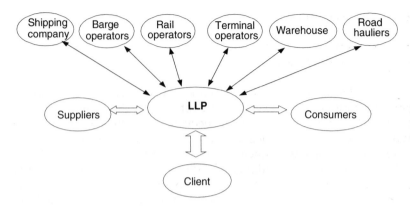

Figure 9.1 The position of LLP in transport logistics

economies of scope as well as scale. A comparison made by Laarhoven et al. (2000) between two surveys, which were carried out separately in 1993 and 1998, shows that there had been a significant increase of outsourcing logistics activities in the shippers who participated in the two surveys. According to the research survey undertaken by Lieb and Randall (1996) some 60 per cent of Fortune 500 companies report having at least one contract with a third-party logistics provider.

Increasing outsourcing provides the probability of the concentration of logistics service both in industry and at a location. A hub port may take advantage of this process as the gateway position of major seaports offers an opportunity for the enhancement of value-added logistics (Notteboom, 2001). The impact of this process on the quality of a hub port has two main aspects. First, it poses challenges to the capacity of logistics services in a hub port due to the surge of logistics activities. Since the customers' purpose in outsourcing is to gain the expertise and advanced facilities that otherwise would be difficult to acquire, or costly to have in-house (Razzaque and Sheng, 1998). It may be equally problematic for the logistics community in the hub port to acquire the necessary experts, professionals and advanced facilities. Secondly, the increase of outsourcing fosters 3PLs and causes greater specialization in the logistics services industry that is characterized by the emergence of even fourth-party logistics providers (4PLs) or leading logistics providers (LLPs) which have assumed the responsibility for managing all warehousing, transportation, order management and returned goods for a specific client through synchronizing the technology and forming networks and alliances between other 3PLs (Lalonde, 2001). Thus, the LLPs have gained a special position in transport logistics as shown in Figure 9.1, in which the LLP is supposed to play an important role in port selection; consequently, the logistics function in the quality of a hub port will be stressed.

Improvement and integration of logistics services

From JIT (just-in-time) in the early 1970s to the latest collaborative planning, forecasting and Replenishment (CPFR) the trend in supply chain business models is the integration of organizations from upstream to downstream. Its influence on the transport logistics sector is that the customer becomes more demanding in terms of service quality and scope. Shipments become more frequent, fewer in volume and more stringent with regard to delivery time. All of these lead to integration in transport logistics, since successfully integrated logistics management ties all logistics activities together in a system, which simultaneously works to minimize total distribution costs and maintain desired customer service levels (Kenderdine and Larson, 1988; Gustin et al., 1995). Through integration, the 3PLs not only extract costs and efficiency but also find more chances to deliver 'value' to the end customer and gain competitive advantage (Robinson, 2001). This trend is also reflected by customer demand, as shown by the survey carried out by *Containerization International* (2001) on global shippers, where about 39 per cent of total respondents said they were in favour of a 'one-stop' shop, and required their logistics partners to provide warehousing and secondary distribution services as well.

The influence of logistics integration on the quality of hub ports can be summarized in three points:

- First, it poses a challenge to port operations, because there still exist some gaps between organizations in the maritime sector. Avery (2000) points out that a port needs to do more in cooperation with carriers' different customer service strategies. For example, a port should stow the import containers in a way concordant with customers' demand on timeliness.
- Secondly, land accessibility will be emphasized in the quality of a hub port. Considering a large-scale hinterland and long-distance land transportation, the gain in high-quality liner services may be offset by poor intermodal operations and inland transportation. The cost and time consumed in intermodal operations will be a determining factor in a hub port's penetration capability into the inland market. Figure 9.2 shows the cost structures of unimodal road haulage versus multimodal transport, in which the intermodal operational costs are the crucial factor in total transport costs. Thus, the following three aspects will be taken into consideration in evaluating land accessibility and regarded as an important quality of a hub port:

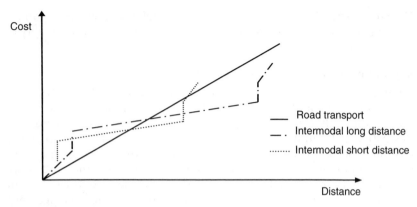

Figure 9.2   Cost structures of unimodal versus multimodal transport
Source: Rutten (1995).

- Institution (e.g. inter-governmental cooperation, custom clearance procedures)
- Operations (e.g. intermodal cooperation) and
- Organization (e.g. the scale and scope of intermodal transportation organizations)
- Finally, the integration of logistics also changes the structure of the maritime community. The introduction of integrated transport practices such as door-to-door or point-to-point transport shifts the port choice from the shipper to the shipping lines (Marti, 1988). The major port clients consider a port merely as a sub-system in the logistics chain; thus port selection becomes merely a function of network costs (Notteboom and Winkelmans, 2001). ⌊⊃ Pon : ≠ pore ⊕igu

Intensive application of information technology

It is said that only about 20 per cent of the door-to-door transport costs of moving goods is spent on actual transportation, with the rest being spent on physical interfaces and on 'soft' services, such as information management and transaction costs (Frankel, 2001). Various authors (e.g. Kerr, 1989; Bowersox and Daugherty, 1995; Fabbe-costes, 1997; Alshawi, 2001) promote information technology (IT) as a means of enhancing logistics competitiveness. The use of IT to form a consolidated logistics network has become an inevitable trend and IT has been regarded as a logistics resource, as well as a competitive weapon (Closs et al., 1997). Some new concepts, such as 'Virtual Logistics', which creates new possibilities in the design of logistics systems are entirely based on IT. According to Berglund et al. (1999), IT skills are at the base of all four of the value creation methods in logistics, which include operational efficiency, sharing resources, the development of a network of service providers, and the use of conceptual logistics skills to improve the customers' supply chains.

In a hub port, IT plays a crucial role in the terminal operation, the distribution centre operation and intermodal cooperation. So it should be regarded as a new dimension of hub port quality, which may include two factors – the level of development and that of application. Although information infrastructures have been regarded as the key factor that determines the efficiency of maritime and port logistics (Kim et al., 2001), ports still lag behind in the adoption of new IT. Park et al. (2001) argue that from a perspective of supply chain management, existing EDI has a lot of problems because it has only one role: transmitting messages. Besides the existing technological obstacles, there are also some other reasons that hinder the application of advanced IT in the port logistics community. The chairman of the Hong Kong Sea Transport Association states that:

> standardization of IT solution platforms is a very crucial issue in the logistics field. But enterprises in the private sector generally are not willing to develop a system to be shared with other enterprises. Some enterprises even think that for their own benefits, it is better to develop a unique system platform, which is different from others. (*Hong Kong Logistics*, 2001)

Broadening the scope of 3PLs' services

A large part of the value creation in the supply chain is transferred to logistics service providers. Value-added logistics might even include secondary manufacturing activities such as systems assembly, testing, software installation, etc. (Notteboom and Winkelmans, 2001). For example, over the last few years, in the automotive industry, sequencing, kitting and subassembly are frequently undertaken at the supplier location or at an additional, intermediate facility, which is run by the supplier or a third-party logistics provider (Mathisson and Johansson, 2000). According to a survey conducted in Denmark by Larson and Gammelgaard (2001), there are some relatively new services that are provided by 3PLs, such as supply chain design, cross-docking operation, assembly and production.

As a result, the port turns into a service centre, through which port activities are diversified from pure cargo handling at terminal points to a comprehensive transport service package (Stuchtey, 1990). Functions such as a distribution centre have played an important role in this aspect. Although the distribution centre is not a new concept, as it has existed in hub ports such as Hong Kong for decades, the service quality and scope in the distribution centre has been continuously improved and broadened. Table 9.3 shows the main service provided by current distribution centres. The location of a distribution centre or 'Distripark' has been frequently taken into account when designing new terminals in a port to achieve the most efficient and lowest cost of distribution service, and a location close to a port has been considered to be a significant advantage to a distribution centre.

Table 9.3   Services of a distribution centre

Basic Activities
Loading and unloading.
Distribution and consolidation of cargo.
Unitization and deunitization.
Storage.
Other Value Adding Activities
Packing.
Testing.
Information processing.
Sorting and sequencing.
Assembling.
Marking.
Customs clearance.
Administrative activities.
Inventory control.
Exhibition.

Source: International Association of Ports and Harbours (1996).

In summary, throughout logistics development, various forms of challenge have been posed to a hub port as it is gradually converted into a logistics platform. To meet these challenges, new dimensions and/or new factors should be identified as affecting the quality of hub ports. The following section will develop a new framework for the new era as a new paradigm of hub ports.

9.5   A new paradigm of hub port quality in the logistics era

This section synthesizes the dimensions and/or factors affecting hub port quality, together with the influence of logistics development on its quality that have been identified in the previous sections. A conceptual framework will be established for the quality of hub ports, which is more pragmatic

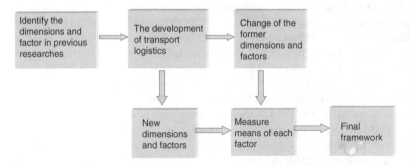

Figure 9.3   The procedure for the framework of hub port quality

in examining the current and potential hub ports. The whole process is illustrated in Figure 9.3.

The framework has three key parts:

- The dimensions of the quality of hub port.
- The factors of each dimension; and
- The measuring means of each factor, as shown in Table 9.4.

Table 9.4   A new dimension and factor for hub port quality in the logistics era

Dimension	Factors	Measure means
Port location	Distance to the industrial agglomeration region	Referencing geographical information
	Distance to the mainlines Strategic location in the global network	Surveying carriers and LLPs
Port infrastructure and superstructure	Berth number Berth depth Crane type Yard area	The requirements of accommodating the latest generation container ship
Port service	Load and discharge speed Pick up and delivery service Information availability Provide customized service Ancillary service	Referencing to the international benchmarks Surveying shippers and forwarders Surveying carriers
Port charge and cost	Port charge of cargo Port charge of ship	Referencing to the international benchmarks
Carrier's service in port	The calling frequency The freight rate Carriers' participation of inland transport	Referencing the published data Referencing the published data Surveying shippers and LLPs
Hinterland accessibility	Intermodal operation time consume Intermodal operation cost Custom clearance procedure Cargo tracing serve	Referencing to the international benchmarks Surveying carriers and LLPs
Distribution centre (DC)	Total operation area of DC The equipment and information system of DC Service scope	Referencing statistic data Requirements of the advanced supply chain management (Surveying LLPs)
Info-structure	Port community system Information interchange with custom Information exchange between the intermodal organizations	Requirements of the advance supply chain management (Surveying LLPs)

The final dimensions include those identified in previous research and the ones appearing in the logistics era (i.e. distribution centres and info-structure). All of the dimensions are also reassessed in the context of logistics development, and necessary adjustments have been made on the factors they contained. In addition, the measuring tools for each factor have been suggested as a guide for further empirical study. A survey methodology is proposed for the factors that are difficult to quantify. As they gradually play a bigger role in port selection as shown in Figure 9.1, LLPs are added to shippers, carriers and forwarders as the main appraisers of various factors and the respondents to the survey.

9.6   Conclusions

The quality of hub ports has become a more sophisticated concept in the logistics era. New dimensions such as distribution centres and info-structure appear to be increasingly important, and traditional functions need to meet the new requirements of logistics services. This chapter conceptually constructs the full dimensions for assessing the quality of a hub port in the new era that we find ourselves in. As an area for further investigation, an empirical study is to be conducted, and more research should be carried out on single- or multi-dimensions and/or factors such as spatial factors, info-structure, intermodality or distribution centres in order to find out how and to what extent they contribute to hub ports. Finally, policy- and regulation-related research should be carried out in conjunction with the aforementioned dimensions in order to see how logistics development in hub ports is spurred on through policy and regulations.

References

Alderton, P. (1995) *Sea Transport: Operation and Economics*, London: Thomas Reed Publications.

Alderton, P. (1999) *Port Management and Operations*, London: LLP.

Alshawi, S. (2001) Logistics in the Internet Age: Towards a Holistic Information and Processes Picture, *Logistics Information Management*, 14(4), 235–42.

Avery, P. (2000) *Strategies for Container Ports*, London: Cargo Systems.

Berglund, M. and Laarhoven, P. and Sharman, G. and Wandel, S. (1999) Third-Party Logistics: Is There a Future?, *International Journal of Logistics Management*, 10(1), 59–77.

Bowersox, D. and Daugherty, P. (1995) Logistics Paradigms: the Impact of Information Technology, *Journal of Business Logistics*, 16(1), 65–80.

Closs, D., Goldsby, T. and Clinton, S. (1997) Information Technology Influences on World Class Logistics Capability, *International Journal of Physical Distribution and Logistics Management*, 27(1), 4–17.

Containerization International (2001) Take your Partners, November, 57–61.

Council of Logistics Management (1998) *What it's All About*, Oak Brook, IL: Council of Logistics Management.

Cullinane, K. (2000) The Competitive Position of the Port of Hong Kong, *The Journal of the Korean Association of Shipping Studies*, 31, 45–61.

Fabbe-costes, N. (1997) Information Management in the Logistics Service Industry: A Strategic Response to the Reintegration of Logistical Activities, *Transport Logistics*, 1(2), 115–27.

Fleming, D. and Hayuth, Y. (1994) Spatial Characteristics of Transportation Hubs: Centrality and Intermediacy, *Journal of Transport Geography*, 2(1), 3–18.

Foster, T. (1978) What's Important in a Port, *Distribution Worldwide*, 78(1), 34.

Frankel, E. (2001) The New World of E-Logistics and Shipping, *Singapore Maritime and Port Journal*, 3, 48–54.

Gem, K. (1998) *Transforming Korea into a Logistics Centre for Northeast Asia*, Anyang: Korea Research Institute for Human Settlements.

Gustin, C., Daugherty, P. and Stank, T. (1995) The Effects of Information Availability on Logistics Integration, *Journal of Business Logistics*, 16(1), 1–21.

Hayuth, Y. (1978) *Containerization and the Load Centre Concept*, PhD Dissertation (Seattle: University of Washington).

Hayuth, Y. (1987) *Intermodality: Concept and Practice*, London: LLP.

Hayuth, Y. (1995) Container Traffic in Ocean Shipping Policy, *International Conference Ports for Europe*, Zeehaven Brugge: Brugge.

Hayuth, Y. and Fleming, D. (1994) Concepts of Strategic Commercial Location: the Case of Container Ports, *Maritime Policy and Management*, 21(3), 187–93.

Hong Kong Logistics (2001) The Lack of Standardization of IT Solution Platforms – A Problem Facing the Logistic Community, *Hong Kong Logistics*, 4(1), 22.

Hormer, M. and O'Kelly, M. (2001) Embedding Economies of Scale Concepts for Hub Network Design, *Journal of Transport Geography*, 9(4), 255–65.

International Association of Ports and Harbours (1996) *The Future Role of Ports in Combined Transport and Distribution Centre*, Tokyo: IAPH.

Kenderdine, J. and Larson, P. (1988) Quality and Logistics: a Framework for Strategic Integration, *International Journal of Physical Distribution and Materials Management*, 18(7), 5–10.

Kerr, A. (1989) Information Technology – Creating Strategic Opportunities for Logistics, *International Journal of Physical Distribution and Logistics Management*, 19(5), 15–17.

Kim, H., Park, N., Choi, H. and Kim, S. (2001) Development of XML/EDI Based Data Warehouse for Customs Clearance of Maritime Exports, IAME Annual Conference 2001, Hong Kong.

Laarhoven, P., Berglund, M. and Peters, M. (2000) Third-party Logistics in Europe – Five Years Later, *International Journal of Physical Distribution and Logistics Management*, 30(5), 425–42.

Lalonde, B. (2001) New Roles and Risks for 3PLs, *Supply Chain*, September/October, 9–10.

Larson, P. and Gammelgaard, B. (2001) Logistics in Denmark: A Survey of the Industry, *International Journal of Logistics*, 4(2), 191–206.

Lieb, R.C. and Randall, H.L. (1996) A Comparison of the Use of Third-party Logistics Services by Large American Manufacturers, 1991, 1994, and 1995, *Journal of Business Logistics*, 17(1), 305–20.

Lloyd's List Maritime Asia (2001) Ever Gone, November, 6–7.

Malchow, M. and Kanafani, A. (2001) A Disaggregate Analysis of Factors Influencing Port Selection, *Maritime Policy and Management*, 28(3), 265–77.

Marti, B. (1988) The Evolution of Pacific Basin Load Centres, *Maritime Policy and Management*, 15(1), 57–66.

Mathisson, B. and Johansson, M. (2000) Influences of Process Location on Materials Handling: Case From the Automotive Industry, *International Journal of Logistics: Research and Application*, 3(1), 26–39.

Murphy, P., Dalenberg, D. and Daley, J. (1988) A Contemporary Perspective on International Port Operations, *Transportation Journal*, 28(1), 23–32.

Murphy, P., Dalenberg, D. and Daley, J. (1989) Assessing International Port Operations, *International Journal of Physical Distribution and Materials Management*, 19(9), 3–10.

Murphy, P., Dalenberg, D. and Daley, J. (1991) Analyzing International Water Transportation: The Perspectives of Large US Industrial Corporations, *Journal of Business Logistics*, 12(1), 169–90.

Murphy, P., Daley, J. and Dalenberg, D (1992) Port Selection Criteria: An Application of a Transportation Research Framework, *Logistics and Transportation Review*, 28(3), 237–55.

Notteboom, T. (1997) Concentration and Load Centre Development in the European Container Port System, *Journal of Transport Geography*, 5(2), 99–115.

Notteboom, T. (2001) Spatial and Functional Integration of Container Port Systems and Hinterland Networks in Europe, *Report of the 113 Round Table on Transport Economics: Land Access to Sea Ports*, Paris: OECD.

Notteboom, T. and Winkelmans, W. (1998) Spatial Concentration of Container Flows: the Development of Load Centre Ports and Inland Hubs in Europe, 8th WCTR Conference, Antwerp.

Notteboom, T. and Winkelmans, W. (2001) Structural Change in Logistics: How Will Port Authorities Face the Challenge?, *Maritime Policy and Management*, 28(1), 71–89.

O'Mahony, H. (1998) *Opportunities for Container Ports: a Cargo Systems Report*, London: Cargo Systems.

Park, N., Kim, H., Choi, H. and Cho, J. (2001) One-stop System Model for Port and Logistics Using SCM, IAME Annual Conference 2001, Hong Kong.

Rabinovich, E., Windle, R., Dresner, M. and Corsi, T. (1999) Outsourcing of Integrated Logistics Functions an Examination of Industry Practices, *International Journal of Physical Distribution and Logistics Management*, 29(6), 353–73.

Razzaque, M. and Sheng, C. (1998) Outsourcing of Logistics Functions: a Literature Survey, *International Journal of Physical Distribution and Logistics Management*, 28(2), 89–107.

Robinson, R. (1998) Asian Hub/Feeder Nets: the Dynamics of Restructuring, *Maritime Policy and Management*, 25(1), 21–40.

Robinson, R. (2001) Ports as Elements in Value-driven Chain Systems: the New Paradigm, IAME Conference 2001, Hong Kong.

Rutten, B. (1995) On Medium Distance Intermodal Rail Transport, MSc Dissertation, Delft University of Technology.

Scotti, A. (1998) Planning a High Density Hub Terminal, in R. Mina and M. Kraman (eds) *Ports' 98*, Reston, VA: American Society of Civil Engineers.

Slack, B. (1985) Containerization, Inter-port Competition, and Port Selection, *Maritime Policy and Management*, 12(4), 297–9.

Slack, B. (1994) Domestic Containerization and the Load Centre Concept, *Maritime Policy and Management*, 21(3), 229–36.

Starr, J. (1994) The Mid-Atlantic Load Centre: Baltimore or Hampton Roads?, *Maritime Policy and Management*, 21(3), 219–27.

Stuchtey, R. (1990) *Port Management Textbook*, Bremen: Institute of Shipping Economics and Logistics.

Taaffe, E., Gauthier, H. and O'Kelly, M. (1996) *Geography of Transportation*, London: Prentice-Hall International.

Tongzon, J. (2001) Key Success Factors for Transhipment Hubs: The Case of the Port of Singapore, IAME Annual Conference 2001, Hong Kong.

Wang, J. (1998) A Container Load Centre with a Developing Hinterland: A Case Study of Hong Kong, *Journal of Transport Geography*, 6(3), 187–201.

10
Key Success Factors for Transhipment Hubs: The Case of the Port of Singapore
Jose Tongzon

10.1 Introduction

The port of Singapore has an enviable position as one of the busiest container ports in the world. For many years the port of Singapore was lagging behind the port of Hong Kong in terms of container throughput, but in 1998, in the midst of the regional economic crisis, its container throughput grew by about 7 per cent to 15.3 million TEUs, surpassing that achieved by the port of Hong Kong. In terms of shipping tonnage, since 1992 the port of Singapore has consistently ranked as the world's busiest port. In 2001 there were 146,265 ships that called at the port of Singapore, equivalent to 910.1 million gross registered tonnes (GRT). At any one time, 800 ships are waiting to berth at the port. It is also the largest bunkering port in the world, with 18 million tonnes of bunkers sold in 1998. In addition, throughout the 1990s the port of Singapore was consistently voted the Best Container Terminal in Asia (*The Straits Times*, 1999: 57). In the national context, the port has been a major contributor to national income and employment, and, despite the current economic crisis, has shown a certain degree of resilience and determination to maintain its premier port status.

A large proportion of its container traffic flows are transhipments, that is, cargoes originating from Singapore's neighbouring countries or destined for countries outside of Singapore. This is not surprising given the limited size of its domestic market. It has been reported that about 50 per cent of Malaysian sea cargo goes via the port of Singapore (*The Business Times Shipping Times*, 16 October 1997). In dollar terms, some RM164 billion worth of goods are dispatched annually by Malaysian exporters through the port of Singapore (*The New Straits Times*, 21 July 1998).

In 1990, approximately 70 per cent of its four million TEUs was accounted for by containers transhipped from neighbouring ports (Airriess, 1993: 34). Currently more than two-thirds of Singapore's container traffic is transhipped. Large ships do not have to call at every port within the region, but only call at the port of Singapore to unload into, or load cargoes from,

Table 10.1 Proportion of transhipments to total throughputs, in major ports

Ports	Transhipments as percentage of total throughput, 1996
1. Malta	92
2. Algeciras	90
3. Damietta	90
4. Singapore	78
5.Kingston	75
6. Colombo	72
7. Gioia Tauro	65
8. Dubai	48
9. Kaohsiung	43
10. Rotterdam	40
11. Antwerp	35
12. Hamburg	35
13. Pusan	30
14. Felixstowe	28
15. Bremerhaven	25
16. Hong Kong	20
17. Kobe	15

Source: Hoffmann. (2000), Table 1.

feeder ships. This will enable the shipping lines not only to cut their costs, but also to provide a high frequency of service, maximize the utilization of slots of the mother vessels and to have a broad choice of feeders. From the viewpoint of the shippers, this system allows them a wider choice of shipping lines and times at more competitive rates.

To give some indication of the extent to which the port of Singapore relies on transhipments, Table 10.1 shows the proportion of transhipments to total throughputs in selected major ports based on 1996 data. As can be gleaned from the table, Singapore's proportion of transhipment came fourth highest after the ports of Malta, Algeciras, and Damietta, but it is certainly the highest among the major transhipment hubs such as the ports of Hong Kong, Rotterdam, Kaohsiung and Felixstowe.

The role of the port of Singapore largely reflects the nature of its economy, since it has served as a link between its geographical neighbours and their trading partners since colonial times. Given its good geographical location and having made a headstart in infrastructure development relative to its Southeast Asian neighbours, Singapore had the qualities required to play an entrepôt role. Before the second half of the 1980s, the geographical concentration of container shipping services was between Northeast Asia and the United States, with East Coast services transiting the Panama Canal. From this period, when Southeast Asia started to experience remarkable economic growth largely aided by the relocation of investments and manufacturing activities from its Northeast Asian neighbours, the role of the port of Singapore as the region's hub port has further strengthened. Major shipping consortia began to introduce services that extended westwards to Singapore, transhipping cargoes to and from other Southeast Asian ports over that port.

What are the key factors responsible for the success of the port of Singapore as a transhipment hub? What role did the government policy makers and the Port of Singapore Authority in particular play in its success? What are the potential threats to its transhipment business in the midst of globalization and growing inter-port competition, and what policies are being adopted to deal with these threats? This chapter tries to answer these questions with the intention of drawing some general lessons for other ports aspiring to become future transhipment hubs.

10.2   Key success factors

A number of factors have contributed to the success of the port of Singapore as a transhipment hub, but the following are considered to be crucial:

- Its strategic location.•
- Its high level of operational efficiency.
- Its high port connectivity.
- Its adequate infrastructure.
- Its adequate info-structure and a wide range of port services.

Strategic location

To become a hub port, a port must be strategically located. A port is considered to be strategically located if it has at least one of the following three characteristics: situated on the main maritime routes; situated in or near production and/or consumption centres; possessing natural deep water harbours, natural breakwater and big waterfront and landside development possibilities.[1] Moreover, a good geographical location should also be one where favourable climatic conditions prevail. Harsh weather can obstruct the daily operations at a port and hinder its development.[2]

Fleming and Hayuth (1994) have stressed the importance of location advantages to load centring, and Fleming (1997) has further observed that the world's top container ports are endowed with such location attributes as centrality and intermediacy, in various and varying proportions. They are all central to something and en route to something else, which is what one might expect of any large seaport gateway. Hong Kong, for example, has been chosen by major container lines as an interchange point between mainline and feeder services. Hong Kong's close connections with the Pearl River Delta and the booming southern province of China generate a lot of traffic. Although Hong Kong accounts for only about 20 per cent of their

TEU total as true transhipments, moving on through bills of lading, there are more re-exports and consolidations in Hong Kong and up to 70 per cent of the TEU total either originates in or ends up in China (Fleming, 1997: 178).

The port of Singapore is located along the Straits of Malacca, which is a main shipping route between East and West. It was recently estimated that over 600 ships transit the straits every day (*The Business Times Shipping Times*, October 1997: 1).

It is also fortunate in enjoying natural deep waters and harbours, which allow it to service ships with deeper draughts without necessarily resorting to extensive and expensive dredging operations. The waterways giving access to Singapore allow even the largest ships to use them. Singapore also does not suffer from typhoons and other natural calamities, and this makes port operations and freight movements safe and reliable.

Singapore is located close to some of the world's dynamic economies. Even before rapid economic development of these economies began, the port of Singapore had already played the role as an *entrepôt* port, serving as a gateway to Singapore's hinterland. The recent remarkable economic development and growing trade orientation of its close Asian neighbours have only heightened its role as an entrepôt port. Although the 1997 economic crisis has adversely affected its port business (for example, it witnessed a fall in cargo tonnage of 4–7 per cent in 1998), the long-term future for the region is bright and will remain one of the most dynamic regions in Asia.

High level of port efficiency

Although the geographical location is a prime factor for a hub status, it is worth noting that many ports without such good natural conditions have obtained very large market shares by promoting other competitive factors. In Europe, for example, where many of the largest traditional ports have faced many difficulties in adapting their management structures and labour relations to the needs of the carriers, opportunities for growth have been exploited by smaller ports. Two private ports, Felixstowe and La Spezia, are now the largest container ports in the UK and Italy, respectively. They have eclipsed the former giants London and Genoa, which had become beset with chronic labour problems and planning difficulties (Slack et al., 1996: 298).

Schlie (1996) defined competitiveness as the ability to get customers to choose a particular service over competing alternatives on a *sustainable* basis. In the context of sustainability, ports should think long term and invest for the future, even at the expense of short-term profits. Port efficiency often means *speed* and *reliability* of port services. In a survey conducted for UNCTAD (1992), 'on-time delivery' was cited to be a major concern by most shippers. In fast-paced industries where products must be moved to the markets on time, terminal operators as vital nodes in the logistics chain must be in a position to guarantee shipping lines very reliable service levels. These include the on-time berthing of vessels, guaranteed turnaround time[3] for vessels and guaranteed connection of containers.

The turnaround time of ships can be reflected in the freight rates charged by shipping companies and cargo dwelling time. The longer a ship stays at berth, the higher is the cost that a ship will have to pay. This can be passed on to shippers in terms of higher freight charges and longer cargo dwelling time, and thus reduce the attractiveness for them to hub at a port. Tongzon and Ganesalingam (1994) identified *several* indicators of port efficiency and categorized them into two broad groups: namely, operational efficiency measures and customer-oriented measures. The first set of measures deals with capital and labour productivity[4] as well as asset utilization rates.[5] The second set includes direct charges, ship's waiting time, minimization of delays in inland transport and reliability.

Perceptions about port efficiency are often as important as the actual levels of port efficiency. A record of accomplishment based on the number of awards received gives assurance to customers in terms of quality and reliability. The latter is crucial in influencing carrier's choice of hub port, as it is often the relative perception of customers (classified here as water carriers, freight forwarders, larger shippers and smaller shippers) that carries greater weight than the actual port performance.

Port efficiency can be measured in many ways depending on which aspects of port operations are being evaluated. It can also be viewed from different perspectives, as there are a number of port users with their different and, at times, conflicting objectives. For instance, a port may be efficient to shipowners, but may at the same be judged inadequate to cargo interests. Thus, any measure of port efficiency cannot be based solely on one criterion. A meaningful evaluation of a port's efficiency will require a set of indicators relating to the various aspects of port operations. There is, therefore, a need to come up with an aggregate or overall index of port efficiency. To show the overall ranking of the port of Singapore in terms of efficiency, Tongzon (1999) used a mathematical programming technique called data envelopment analysis (DEA) to measure the port of Singapore's overall level of efficiency compared to other major ports in the world. The findings of this study, shown in Table 10.2, have confirmed the port of Singapore's high level of efficiency relative to all other ports in the sample. In all cases, the port of Singapore is ranked as efficient and has appeared in all facets for port efficiency comparison.[6] The port of Klang is also found to be relatively efficient. However, it should be pointed out that this port does not appear in all facets for port efficiency comparison.

Tongzon and Genesalingam (1994) and Tongzon (1995) have also shown the port of Singapore to be in the same league (similar in contexts) as the

Table 10.2 Relative efficiency rankings using DEA

Ports	Additive	CCR
1. Melbourne	0.6954	0.4383
2. Sydney	1.0000	1.0000
3. Brisbane	1.0000	1.0000
4. Fremantle	1.1820	0.2747
5. Adelaide	1.000	0.7564
6. Rotterdam	0.7116	0.5697
7. Tacoma	1.0000	1.0000
8. Zeebrugge	1.0000	1.0000
9. Wellington	1.0000	0.3058
10. Montreal	0.6767	0.3871
11. Baltimore	0.6000	0.3075
12. Auckland	1.0000	0.8446
13. Le Havre	1.0000	0.4219
14. Hong Kong	1.0000	1.0000
15. Kaohsiung	0.7939	0.6412
16. Felixstowe	1.0000	1.0000
17. Puerto Rico	1.0000	0.9944
18. Jakarta	1.0000	0.9157
19. Manila	1.0000	0.3792
20. Klang	1.0000	1.0000
21. Singapore	1.0000	1.0000
22. Bangkok	1.0000	1.0000

Notes: CCR is the Charnes, Cooper and Rhodes (1978) DEA method. Additive is the additive DEA model of Charnes et al. (1985). A score of 1 indicates that the port is efficient.

ports of Rotterdam, Hong Kong and Kaohsiung, and that the port of Singapore has outshone all other similar ports in the area of ship turn-around time, labour efficiency, crane efficiency and in the utilization of other port assets. The port's high level of efficiency has made it more economical for shipping lines to use.

However, the analysis has shown, based on more recent data, that other ports are eroding its advantages as far as efficiency, timeliness and reliability are concerned. For example, the port of Klang has improved its average container-handling rate per ship hour from 35.4 TEUs in 1995 to 51.3 TEUs in 1998. Correspondingly, it has also showed an improvement in respect of ship turnaround time from 12.5 hours in 1995 to 11 hours in 1998.

High port connectivity

A recent transhipment study by UNCTAD (1992) proposes that veritable grid networks, assembled around transhipment ports where different trade routes intersect and interconnect, have replaced traditional port-to-port routes. Big trans-oceanic shipping lines begin to take advantage of the flexibility and the

Table 10.3   Like for like comparative performance of Singapore based on selected indicators, 2000

	Port charges[a] (US$)	Ship turnaround time (hours)	Connectivity to other ports
Port of Singapore	155.0	12	740
Port of Klang	50.0	12.5	500
Port of Tanjung Pelepas	–	–	2 major shipping lines
Port of Bangkok	23.37	–	–
Port of Tanjung Priok	–	–	–
Port of Manila	24.74	–	–
Port of Rotterdam	–	23	1,000
Port of Melbourne	23.28	–	200
Port of Auckland	26.52	14.9	160
Port of Felixstowe	1,051	–	365

Notes: — not available; Ranks from 5 (highest) to 1 (lowest); [a] Represented by container handling rates per FCL; Exchange rates used: US$1 = S$1.74, US$1 = RM3.80, US$1 = THB42.8, US$1 = 450 pesos; US$0.49 = AUD$1; US$1 = NZ$ 2.47; US$1 = 0.6912 pounds.
Source: Derived from *Fairplay Port Guide 1999/2000*; www.cosco.com.au/ports.htm.

scope for modulation allowed by the container technique to reorganize and restructure shipping services to regions of heavy traffic. A hub port should, therefore, provide comprehensive connectivity to other ports.

Time is also of the essence. Containers awaiting transhipment at hub ports are costly and counter-competitive in terms of transit time. Whenever possible, operators should strive for tight connection between feeder and mother vessels. Hub ports should maintain the same objectives and encourage port users who support such initiatives with attractive incentives. To ensure fast connections, hub ports which employ a single, or common user, terminal will have an advantage over those that require inter-terminal transfers (Lim, 1996).

As the trend towards large vessels continues, operators will require a network of connecting feeder vessels to aid in the transportation of cargo to and from other shallow-sea ports in the neighbouring area. A port that provides exhaustive and fast connectivity to other ports is capable of assuming the role of hub port for a defined region.

Four hundred shipping lines to practically 740 ports worldwide link the port. Practically all the major international carriers and shipping lines, 400 of them, call at Singapore. As can be gleaned from Table 10.3 its port connectivity covers all parts of the globe with concentration in Southeast Asia. This wide-ranging port connectivity allows shipping lines to maximize slot utilization on their mother vessels by offering more choice of feeders to various trade routes.[7] Shippers are also able to move their products to/from the markets faster and at lower inventory costs.

Reduction in connection times is made possible by the introduction of Fast Connect. Hailed as the new transhipment system, its main aim is to streamline the still cumbersome procedures involved in shipping tranship-ment containers while making possible tighter connection times within the terminals. Thus, in addition to providing comprehensive connectivity, the port is able to expedite the connections, which reduce a ship's stay at berth, leading to lower charges to shippers.

Adequate infrastructure

Infrastructure defines a port's capacity to handle vessels and container flows. It is generally divided into physical and soft elements. Physical infrastructure includes not only the operational facilities (such as the number of berths, the number of cranes, yards and tugs, and the area of storage space), but also the intermodal transport[8] (such as roads and railways). The soft infrastructure refers to the manpower employed. Maximum deployment of both types will assist in reducing vessel turn-around, thereby increasing the port's capacity to accommodate more vessels and container flows.

The trend towards the deployment of bigger ships and containers made for larger transport units in overseas container transport (at the time of writing, ships with carrying capacity of 8,000 TEUs were only under con-sideration) that require new and capital-intensive transhipment facilities in ports. In some ports, the volumes handled far exceed their planned capaci-ties. This results in port congestion and stifles efficiency and trade growth. Such ports have to incorporate plans for 'safety value' facilities outside the port such as depots, bonded warehouses, linked by road or rail to the port to accommodate 'overflows' of boxes. *capacity management*

In particular, intermodality is essential for the speedy transport of cargoes into and out of a hub port. In Europe, for instance, policy makers have embarked on the Trans-European Networks (TENs) to improve inter-modal transport. Without proper linkages, the efficiency of a port may decline as a result of congestion and delays. Notwithstanding the indirect costs incurred, the reputation of a port is also at stake. Additionally, it must be stressed that the quality of the hard infrastructure should not be neglected at the expense of quantity.

In addition to hard infrastructure, human resource is increasing in impor-tance. Modern ports providing 24-hour service depend to a great extent on the common efforts of the whole port community. Employees at different levels will have a role to play, which ultimately contributes to the well-being of a port. Moreover, the intensive use of state-of-the-art technology would require more knowledge workers who specialize in information technology. Effective management of the human resources can tap the knowledge embedded in individuals, generating ideas and innovations that enhance the competitiveness of a port.

Shipping lines are attracted to hub at ports with adequate and superior infrastructure because they can entrust the entire task of container handling to the port while paying more attention to other core activities. Moreover, they can also enjoy lower charges because of the absence of bottlenecks at such a port.

The port of Singapore has a well-developed port infrastructure, not only in terms of the number of container berths, cranes and adequate storage facilities, but also in terms of the quality of the cranes, the quality and effectiveness of the port/inter-port information systems, the approach channel provided, the preparedness of port management and a wide range of port-related and ship-related services offered. Table 10.4 compares the port of Singapore with similar ports and other ports in the region in terms of physical infrastructure.

To be a regional hub requires in particular an adequate number of berths and other required port facilities to deal with significant volumes of cargo traffic, high frequencies of ship visits and very large ships. It also requires a well-motivated, skilled and cooperative workforce to handle the high level of coordination required as a hub port. To meet these requirements, the port of Singapore has ensured that its port facilities are adequate to handle future increases in cargo traffic and ship visits in the region by investing heavily in port expansion and upgrading. It has also adopted a remuneration system that encourages high productivity and cooperation, rather than confrontation, from port workers.

The almost completed development of the Pasir Panjang terminal, which opened in 1998 (after completing the first phase of the project), will give Singapore an extra handling capacity of 18 million TEUs. Once this terminal becomes fully operational, the port's total container handling capacity

Table 10.4   Adequacy of port infrastructure: comparative study of selected ports

	Number of container berths	No. of container ship calls	Delays (hours)	Number of along-the-shore cranes
Port of Singapore	37	24,015	2.3	115
Port of Klang	13	4,889	—	31
Port of Tanjung Pelepas	6	—	—	—
Port of Bangkok	20	2,415	—	—
Port of Manila	10	5,463	22.0	19
Port of Tanjung Priok	25	3,239	50.0	10
Port of Rotterdam	30	5,544	1.7	66
Port of Melbourne	12	823	0	16
Port of Auckland	3	2,381	—	7
Port of Felixstowe	13	2,677	0.6	29

Sources: Taken from interviews, and respective ports' publications.

is expected to be roughly 36 million TEUs per annum. In addition, the port's terminals are supported by a number of district parks, which provide more than half a million square metres of warehousing in total. A district park is a large covered warehouse, which provides automated storage facilities. Customers can process their documents, pack and unpack, mark, label and assemble their goods for distribution to other distribution centres.

Adequate information structure

Information technology (IT) will be the binding element in economic processes in the twenty-first century. This has led to the coining of a new term, 'Infostructure', which refers to the hard- and software telecommunication and EDI systems built in the port area (UNCTAD, 1992), the telecommunication facilities,[9] which link the port and the city to the rest of the world.

With increased volumes handled at ports and the complexity of mother-feeder connections at hub ports, there has to be a faster conversion towards automation of all aspects of terminal operations from in/out gate to ship-side. The use of IT domestically and between regional and hub ports is essential, because accurate and timely data exchange is critical to short transit times and high frequency feeders. Besides, the transparency of markets –and thus the power of customers – increases: for example, having access to information systems of carriers, customers can compare tariffs and identify causes of delay.

Likewise, information technology is applied increasingly in the area of logistics. The management of logistics is becoming the management of information flows. By information technology, documents of shipments can be made to arrive before the arrival of the goods themselves, so that the goods flow can be processed more efficiently (Cooper et al., 1992). The availability of almost real-time information means that in-situ inventories and storage costs can be reduced.

To sum up, operational and communication efficiency, accurate records of transactions and invoicing and minimum paper shuffling are necessary ingredients for future growth. Increasingly, vessels choose to hub at ports with good telecommunication networks so that movements of cargo can be tracked and traced easily and the documentation process shortened.

The port of Singapore is fully equipped to handle almost all types of vessels, ranging from 80 TEU barges to third-generation vessels of 6,600 TEU capacity. It is one of the most automated ports in the world and uses the latest information technology in every aspect of administration, planning and operations services. Its electronic data interchange system is used to plan berth allocation, ship towage and yard management. The port has what is called the Computer Integrated Terminal Operations System (CITOS), which drives all the operations within the terminals. All containers relayed through Singapore are systematically sorted out for distribution to the second carriers.

Singapore has capitalized on its comparative strength in the area of information technology. Apart from its present container and vessel automated tracking system, it has recently implemented a system whereby shipping and cargo information can be accessed through the internet. Through the down-sized Windows-based PORTNET, the customers are linked with the port of Singapore and with relevant government agencies such as the Maritime Port Authority, the Trade Development Board and the customs department. This Internet-based system has 52 modules and 400 sub-modules offering a wide range of services, including berth application, submission of import status of cargo and 24-hour tracking of containers. Subscribers pay a one-off set-up fee and a transaction fee each time they access the system.

Advanced cargo-handling equipment, such as fourth-generation quay cranes, double trolley cranes and double-stack trailers, are employed. The management is carrying out additional upgrades to its cargo-handling technology by almost entirely automating its terminal operations. In the new Pasir Panjang terminal mentioned earlier, containers are handled and transported by computer-controlled machines. The port management is currently testing automatic guided vehicles (AGV) capable of navigating autonomously and stacking containers by remote-controlled bridge cranes.

Singapore also wants to build up a strong research and development capability in the port and maritime fields and is currently undertaking a study of highly automated container terminals to handle jumbo container ships.

Wide range of port services

Recent statistics have shown that vessels are often calling at a port for more than two purposes. This finding implies that ports should provide a range of services such as bunkering, pilotage, warehousing, and cold storage. Importantly, they should be *integrated* so that ports can be a one-time stop for shipowners. Besides choosing from a competitive supply, present-day customers also want to individualize their purchases.

The globalization of industries has resulted in flows of materials and information from a multitude of sourcing and manufacturing points to a diversity of markets with specific requirements in terms of customer service. Consequently, the routing of logistics chains is becoming more complex. The integral control of activities is becoming a critical factor for competition (van Klink, 1995).

This 'supply chain management' (Ellram and Cooper, 1990) includes coordinating and streamlining the activities of all firms in the chain. A precondition for effective chain integration is the existence of a channel captain, who can bind all parties and activities. With regard to the individualization of the location requirements of activities, the streamlining of operations in logistics chains can result in the spatial separation of activities to relocate every link of the chain at the optimal site. It necessitates the development of centralized warehousing and distribution, which are sometimes linked to final assemblage and marketing (termed 'value-added logistics').

Over time, high performance distribution systems will be required. The systems must be in line with and complement the rest of the shippers' logistics channels, including the material flow at manufacturing plants. Ports can experience synergistic benefits from distribution centres.[10] A distribution centre is advantageous because it attracts cargo that can be shipped through a port. There is a positive relationship between the cargo flow and the ships calling at the port: the cargo attracts the ships and the ships attract the cargo. From all of these activities, the port earns revenue. Hence a port can profit not only from the distribution centre itself, but also from the increased flow of cargo through the port (IAPH, 1996) Thus, an ideal hub port should provide a diverse range of services that are highly integrated. The increasing role of ports in logistics management should be considered seriously. To provide logistics services, the port must have an efficient distribution system.

The port of Singapore has also built a wide range of ship-related and port-related services such as bunkering, ship repair, storage and others. It is now trying to introduce other auxiliary services, such as marine finance, insurance, brokerage, in order to create a one-stop shop centre.

With the combined resources of Alexandra, Pasir Panjang and Tanjong Pagar Distriparks, the port of Singapore offers a total of 462,000 square metres of warehousing space. Located just 15 minutes away from its container terminals and within easy reach of the airport and Pasir Panjang wharves, these Distriparks are home to many established transnational distribution centre operators, manufacturers, traders, forwarders and others, where they enjoy reliable, accessible and well-managed distribution centre operations that are synchronized with their supply chain operations. In addition, its centrally located Keppel Distripark provides the only warehousing facility of more than 10,000 square metres to be found within the Free Trade Zone in Singapore. A wide range of customer-friendly and value-added services such as KD Net and the seamless transfer of cargo to and from the container terminals are also provided, expediting the consolidation of transhipment cargo out of Singapore to the region (Port of Singapore, 1999).

In the mid-1980s Singapore's Trade Development Board (TDB) vigorously promoted Singapore's role as an international warehousing and distribution centre. But the potential of warehousing and distribution was apparent much earlier than this. The Free Trade Zone (FTZ) Advisory Committee considered Singapore as an ideal place for storage and subsequent distribution of goods to the rest of Southeast Asia because of its strategic location and its liberal trading environment. The setting up of free trade zones to facilitate entrepôt trade in dutiable and quota-restricted goods was another factor in securing the success of Singapore as a warehousing and distribution centre. By the end of 1982, there were 6 FTZs in Singapore: Keppel Wharves, Tanjong Pagar Terminal, Jurong Port, Sembawang Wharves, Pasir Panjang Wharves and Changi Airport. In July 1987, Singapore was designated by the London Metal Exchange as the first delivery port outside Europe, a decision which boosted metal trading in Singapore and led to the establishment of several warehouse operations for metals. Further, a number of international companies have set up warehousing and distribution operations, including Eastman Kodak, Caterpillar, Black and Decker, Aiwa, Bayer and Actus. In December 1988, Nedlloyd Districentre established an operation in the Jurong Area. In April 1989, Singapore-based CWT Distribution Pte Ltd. opened its distribution centre – CWT Distripark– considered at that time to be the region's most advanced distribution centre. In more recent developments, Avnet has set up its Asia-Pacific HQ and Distribution Hub to provide Asian contract manufacturers and original equipment manufacturers (OEMs) with services such as procurement, inventory management, just-in-time delivery and design services. Futhermore, Circle International has established a Regional Logistics Centre, which also houses its Asia-Pacific HQ and provides services for the region, including supply chain consulting, materials management, regional distribution, logistics call centre and trade financing. UPS has become the first air express company in Singapore to consolidate air express and logistics operations into a single facility, from which UPS and UPS Logistics will offer integrated logistics services and solutions and also base their Asia-Pacific HQ. DHL has set up its largest express centre in Asia and its largest Regional Air Express Hub in Singapore to provide inventory management, configuration and call centre services. Finally, FedEx has opened its Asia-Pacific Data Centre which also housed its IT and e-commerce development team (Singapore Economic Development Board, 2000).

The concept of a warehousing and distribution centre is strongly supported by the government of Singapore, which sees this to be at least as vital as its cargo-handling function to the promotion of the port as a fully-fledged global hub. With shipping lines increasingly looking to change their modus operandi from being simply carriers of cargo to becoming total transportation specialists, including distribution and warehousing functions, the port of Singapore has considered distribution to be an integral part of its wide-ranging port services.

In summary, there are six key factors responsible for the success of the port of Singapore as a transhipment port:

- Strategic location.
- High level of port efficiency.
- High port connectivity.
- Adequate infrastructure.

- Adequate info structure; and
- A wide range of port services.

However, these factors should not be treated in isolation as they are inter-related and complement each other.

10.3 Role of the Singapore government

Apart from its strategic location, the key factors that have made the port of Singapore a successful transhipment hub did not come about by luck or coincidence. They were the result of proactive government intervention and the effective implementation of appropriate seaport policies. The government of Singapore, in particular, has formulated and effectively implemented seaport policies and strategies to create an environment that nurtures openness, efficiency and accountability in its port operations and services.

Singapore's style of government intervention is a combination of no distortion and *dirigiste*. It encourages competition and operational efficiency by adopting a policy of openness (no import restrictions, price controls or subsidies). On the other hand, it intervenes in the economy both directly (i.e. running government enterprises) and indirectly by way of regulation and other policies affecting the private sector.

The port of Singapore is a good example of this type of government intervention. The port is run on a commercial basis, is self-financing, and is expected to compete with other ports on an equal footing. However, this is also a public port and it is expected to operate with objectives that are consistent with the national development agenda and priorities of the government of Singapore. The Port of Singapore Authority (PSA), operating from 1 April 1964 and later renamed the Maritime Port Authority (MPA) of Singapore following the corporatization of PSA in 1997, is a statutory board whose main task is to regulate and control navigation and shipping in the port area and is under the Ministry of Communications.

The PSA has been responsible for designing and implementing appropriate strategies to meet its objectives. Over the years it has shown its effectiveness in ensuring that the port of Singapore can respond efficiently to new business opportunities and cope with challenges in a changing global trading environment.

For example, in achieving the objective of making Singapore a world-class transhipment port, PSA has adopted a number of incentive schemes, in addition to ensuring that port capacity and infrastructure are adequate to deal with future demand for port services. The 'China Cargo Scheme' was introduced in 1975 to increase trade with China. Similarly, the 'Berth Appropriation Scheme' has been adopted since 1979 for certain major shipping lines. Under the scheme a shipping line is given priority in berthing its vessel at a designated berth and has exclusive use of the 'go-down' behind it, which enables the shipping line to schedule its vessel calls and plan its operations more efficiently. On its part, the shipping line has to guarantee a minimum cargo throughput and to use its own stevedores and gear.

To attract more shipping lines to its port, PSA has adopted a strategy of functional diversification. Apart from the standard role of handling cargoes between the ship and shore, ship repair, bunkering and distribution, other roles such as offering ship finance and insurance have been promoted by the port of Singapore so that it would become a one-stop shop for shipowners. Deregulation in bunkering has also made it easy for foreign companies to establish their own agents in Singapore, instead of having to operate through Singapore agents or foreign-local joint ventures.

It has also promoted the Singapore flag. As a result, the Singapore Registry of Ships has grown rapidly in recent years and today ranks as the seventh largest in the world.

To attract shipowners to set up base in Singapore, since 1991 the Trade and Development Board has implemented an attractive incentive scheme called, the *Approved Shipping Enterprise Scheme* (AIS). Under this scheme companies enjoy a tax exemption of up to ten years on income earned from qualifying shipping operations. To qualify for the scheme, 10 per cent of the company's fleet has to be registered with the Singapore flag. The scheme has proved a strong draw for some of the world's top shipping companies, with 36 AIS companies operating a total of more than 500 vessels using Singapore as a base.

Singapore now has a thriving ship financing community with shipping banks such as ING, Christiana Bank and Mees Pierson joining the local player – the Development Bank of Singapore. Progress has also been made in building up the local insurance industry. Two Protection and Idemnity (P&I) clubs, the Standard Club and the UK Club, now have regional offices in Singapore. Lloyd's of London is continuing negotiations to set up an underwriting presence in the republic.

Foresight and pragmatism also characterize the government of Singapore's policy making. Recently, in the light of growing inter-port competition and the need to consider customer needs, the port of Singapore has adopted a policy of strategic alliances.

Traditionally, Singapore's strategy is aimed at improving its infrastructure, level of efficiency and quality of service, as discussed in the previous section.

Further, it endeavours to increase the profile of the port of Singapore internationally by actively participating in the activities of the International Maritime Organization (IMO) to safeguard its strategic maritime interests and keep the sea safe and open to navigation, and by establishing links with other like-minded countries in the form of inviting certain prominent individuals in international advisory groups and as distinguished visitors. It tries to promote Singapore as a total logistics centre where door-to-door services are available with state-of-the-art logistical facilities and infrastructure and which can add value to the activities of the shipping lines by promoting the bunkering industry, ship registry and other maritime-related activities, including marine finance, insurance, brokerage and others that make up a one-stop centre.

To increase its ability to respond to demands from a fast-changing business environment and deal with increasing threats from other port competitors, it has adopted a policy of corporatization. Since 1997, the corporatized port of Singapore has been able to provide value-for-money services to its customers with even greater speed, quality and reliability. It has been able to maintain and further improve its efficiency by liberalizing its towage services and looking for ways in which competition could be enhanced. It continues to offer major shipping lines priority berthing in exchange for meeting certain conditions – such as meeting a minimum number of containers per year. And it has also been looking for ways to reduce port dues and concentrate on its commercial operations.

Recently, a new approach has been adopted based on a policy of active engagement with other ports. As part of this approach, the port of Singapore has been marketing its consultancy services internationally, particularly in information technology-based port operations and port terminal logistics management. The port of Singapore has also forged certain alliances and cooperative ventures with other ports even as far away as China, India and Africa, offering its capital and expertise in developing and managing state-of-the-art ports. Through these overseas ventures it hopes to build up stronger port linkages with other countries via hub–spoke networks, as well as supplementing its local operations in terms of cargo size.[11]

For instance, in 2001 Singapore handled four million TEUs from its overseas operations and for the year 2002, 8 million TEUs are expected from its overseas operations. To date, in terms of global acquisitions, Singapore now has 13 overseas port projects in ten countries. Thus, in this way it can maintain its position as a global hub port by maintaining its huge cargo base and by having greater influence over the supply lines of transhipment cargo from other ports in the region.

10.4 Conclusions

The success of the port of Singapore as a transhipment hub can be attributed not only to its unique geographical location (along one of the busiest sea lanes), but also to other factors, which have occurred largely as a result of the government's effective, pragmatic and forward-looking policies. Transhipment is a complex operation involving the loading and unloading of specific cargo and its redistribution to ships headed for other destinations.

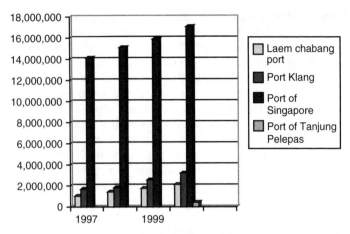

Figure 10.1   Comparison of port throughput, 1997–2000 (TEUs)
Source: Computed from figures taken from official website
(http://www.lcp.pat.or.th), (http://www.psa.com.sg),
(http://www.kct.com.my), (http://www.ptp.com.my) and Fairplay portguide 1999/2000

Singapore's ability to satisfy its customers in terms of efficiency and quality of service has more than compensated for its lack of space and high costs relative to other ports. Although its stevedoring rates are higher than its rivals, the higher levels of efficiency it has achieved have made it more economical for port users to use.

In the face of growing inter-port competition, the port of Singapore has adopted a two-pronged strategy of competition and strategic alliances with other ports. Although other ports in the region are catching up in terms of efficiency and throughput, it is very likely that the premier position of the port of Singapore as a regional hub will remain unchallenged, although its market share may decline, for the following reasons. First, although the other ports are catching up in terms of productivity and infrastructure and thus drawing more main liners to call at their ports, the volumes of cargoes at these ports are still well below those recorded at the port of Singapore. It would take several years before these ports could match the volume achieved by the port of Singapore. As Figure 10.1 shows, the port of Singapore is substantially ahead of other competing ports in terms of throughput, although these ports have experienced steady growth in recent years.

Secondly, the port of Singapore has an advantage as an incumbent. Its world-class reputation as an efficient port with well-established links with major shipping lines and other port users has made it unlikely that the port of Singapore will lose its dominance in the transhipment business in the near future.

Notes

1 Ports without these favourable conditions have to dredge the harbour and build breakwaters, which increase the cost of port services.
2 For instance, some ports are ice-bound in the winter, while in the Middle East, dock working ceases during the midday period in the summer owing to the extreme heat (Branch, 1986).
3 The total turnaround time takes into consideration waiting for pilot, piloting to berth, terminal time, unberthing and final departure from port area.
4 Among the indicators of capital and labour productivity are crane rates (number of containers lifted per net crane hour), ship rates (rates at which cranes load or unload a ship), TEUs per crane (number of containers handled per crane, shipcalls per tug and ship calls per employee. For details, see Tongzon and Ganesalingam (1994).
5 Indicators of asset utilization rates are TEUs per berth metre, berth occupancy and TEUs per hectare of terminal area. See Tongzon and Ganesalingam (1994).
6 For an inefficient decision-making unit (DMU), a facet is a combination of efficient DMUs, which are used to determine the relative inefficiency. The CCR model is based on a constant returns to scale assumption. Under a variable returns to scale assumption, the port of Singapore is also found to be efficient with the score of 1.0000.
7 In terms of connections, at the port of Singapore daily there were 3 sailings to the US, 4 to Japan, 5 to Europe and 22 to South and Southeast Asia at the time of writing.
8 Intermodal transport is a transport of unit loads by the coordinated use of more than one transport mode.
9 Examples include fax machines, IDD (International direct Dial) telephones or even a computerized EDI system, which is linked to the world network.
10 The basic function of a distribution centre is to act as a platform from which to arrange distribution and (de-)consolidate cargo in time and space.
11 The port of Singapore aims to achieve 20 per cent of its annual revenue derived from these overseas ventures, particularly from various strategic alliances and investments in the logistics business and port terminal development. It is currently involved in projects in China, India, Indonesia, Vietnam, South Korea, Hong Kong, Italy and, quite recently, Brunei.

References

Airriess, C. (1993) Export-Oriented Manufacturing and Container Transport, *Geography*, 78(1), 31–42.

Branch, A.E. (1986) *Elements of Port Operation and Management*, London: Chapman & Hall.

Charnes, A., Cooper, W.W. and Rhodes, E. (1978) Measuring the Efficiency of Decision-Making Units, *European Journal of Operational Research*, 2, 429–44.

Charnes, A., Cooper, W.W., Golany, B., Seiford, L. and Stutz, B. (1985) Foundations of Data Envelopment Analysis for Pareto–Koopmans Efficient Empirical Production Functions, *Journal of Econometrics*, 30, 91–107.

Comtois, C. (1994) The Evolution of Containerization in East Asia, *Maritime Policy and Management*, 21(3), 195–205.

Cooper, J., Browne, M. and Peters, M. (1992) *European Logistics – Markets, Management and Strategy*, Oxford: Blackwell Publishers.

Ellram, L.M. and Cooper, M.C. (1990) Supply Chain Management, Partnerships and the Shipper-Third Party Relationship, *International Journal of Logistics Management*, 1(2), 1–10.

ESCAP (1997) *Intra-Regional Container Shipping Study: Prospects for Container Shipping and Port Development*, New York: United Nations Economic and Social Commission for Asia and the Pacific.

Fleming, D.K. (1997) World Container Port Rankings, *Maritime Policy and Management*, 24(2), 175–81.

Fleming, D. and Hayuth, Y. (1994) Spatial Characteristics of Transportation Hubs: Centrality and Intermediacy, *Journal of Transport Geography*, 2(1), 3–18.

Hoffmann, J. (2000) *Concentration in Liner Shipping: Its Causes and Impacts and Shipping Services in Developing Regions*. Transport Unit, International Trade, Financing and Transport Division.

International Association of Ports and Harbours (IAPH) (1996) *The Future Role of Ports in Combined Transport and Distribution Centres*, IAPH Head Office, Tokyo, October.

Lim, D. (1996) Global Ports of the 21st Century, *SingaPort'96 and SibCon'96 Conferences*, Singapore.

Port of Singapore (1999) *The PSA Corporation Annual Report 1999*, Singapore: Port of Singapore.

Singapore Economic Development Board (2000) *SEDB Annual Report 2000*, Singapore: SEDB.

Schlie, T.W. (1996) The contribution of technology to competitive advantage, in H.G. Gaynor (ed.), *Handbook of Technology of Management*, New York: McGraw Hill, 7.1–7.27.

Slack, B., Comtois, C. and Sletmo, G. (1996) Shipping lines as agents of change in the port industry, *Maritime Policy and Management*, 23(3), 289–300.

The Business Times Shipping Times (1997) KL Shippers' Group to Promote Local Ports, 16 October, p. 1.

The Business Times Shipping Times (1997) Accidents in Straits of Malacca up, 16 October, p. 1.

The Straits Times (1999) Changi Airport and PSA are Tops in Region, 15 March, p. 57.

The Straits Times (2000) Portnet has the Edge in Technology, 21 November, p. 14.

The New Straits Times (1998) The Economics of Port Klang, 21 July.

Tongzon, J. and Ganesalingam, S. (1994) Evaluation of ASEAN port Performance and Efficiency, *Asian Economic Journal*, 8(3), 317–30.

Tongzon, J. (1995) Systematizing International Benchmarking for Ports, *Maritime Policy and Management*, 22(2), 171–7.

Tongzon, J. (1999) Measuring Port Efficiency: An Application of Data Envelopment Analysis, Paper presented at the Taipei Conference on Efficiency and Productivity Growth, 30–1 July, Taiwan.

UNCTAD (1992) *Strategic Planning for Port Authorities*, Geneva: United Nations Conference on Trade and Development.

UNCTAD (1996) *Port Organization and Management*, Geneva: United Nations Conference on Trade and Development.

van Klink, H.A. (1995) *Towards the Borderless Main Port Rotterdam: An Analysis of Functional, Spatial and Administrative Dynamics in Port Systems*. Amsterdam: Thesis Publishers.

11
The Logistics Strategy of Japanese Ports: the Case of Kobe and Osaka

Kunio Miyashita

11.1   Introduction

Under intense global competition, ports cannot survive as the nodes of physical distribution without combining efficient cargo flows with appropriate links and concentrating efforts on reducing shippers' total costs.

Ports, in general, are obliged to follow shippers' logistics management strategies. The intent is to make ports operate as just one part of the shippers' logistics systems. Shippers select the ports, rather than the reverse.

At which stage of logistics development do the Japanese ports function as the nodal point of logistics systems? Regrettably, we are currently still in the preliminary stages of logistics development. At present, it is necessary for Japanese ports to concentrate on improving the efficiency of cargo flows by introducing information technology to aid port management. However, in order to accelerate the revolutionary change taking place in Japanese port management systems in an era of deregulation, it is necessary and important to investigate current export and import mechanisms of port behaviour from the viewpoint of trade relations with other Asian countries and economies.

The initial purpose of this research is to conduct an econometric investigation of the current mechanisms of Japanese port behaviour. More concretely, special attention is given to the port of Kobe and the neighbouring port of Osaka. These ports clearly have a competitive relationship and are managed by completely independent port bureaus under the control of the city of Kobe and the city of Osaka respectively. Their competitive relationship can be confirmed by the correlation analysis detailed in Tables 11.1 and 11.2, where the port of Kobe has a negative sign in relation to the correlation coefficient. This indicates that the higher the market share of the port of Osaka, the lower the share of the port of Kobe. This holds true in terms not only of exports but also of imports.

The 'unstable' condition of the port of Kobe is also clearly evident in Tables 11.1 and 11.2. Two other ports (Nagoya and Kitakyushu) are also

Table 11.1   Correlation matrix of the export share of western Japan's four main ports measured by the export volume to nine Asian countries and economies

	Kobe	Osaka	Nagoya
Osaka	-0.406[a]	—	—
Nagoya	-0.320[a]	-0.155	—
Kitakyushu	-0.544[a]	0.091	0.010

Notes: a: Significant at 1% level.
Nine Asian countries and economies: NIES, ASEAN 4, and China.

Table 11.2   Correlation matrix of the import share of western Japan's four main ports measured by the import volume to nine Asian countries and economies

	Kobe	Osaka	Nagoya
Osaka	-0.224[b]	—	—
Nagoya	-0.176	-0.020	—
Kitakyushu	-0.321[a]	0.273	0.131

Notes: a: Significant at 1% level.
b: Significant at 5% level.
Nine Asian countries and economies: NIES, ASEAN 4, and China.

clearly prime competitors to the port of Kobe. In almost all cases, the correlation coefficients between Kobe and the other two ports are significant and negative. In contrast, the port of Osaka has a 'friendly' or complementary relationship with Nagoya and Kitakyushu.

On the basis of this analysis, it is logical to compare the merits and demerits of both the current and hypothetical ('virtual') types of port management systems. In other words, it is important to consider which kind of management system is best able to overcome the existing shortcomings in the port of Kobe and also in the port of Osaka. This is the second purpose of this chapter.

The volume of container ships which called at the port of Kobe in 2000 had almost returned to the same level as before the Great Kobe-Hanshin Earthquake in 1994. However, the level of container cargo volume in 2000 is only 80 per cent of the 1994 base year. According to world container cargo rankings, the port of Kobe is 17th, putting it behind the ports of Tokyo and Yokohama. However, the export container cargo volume through the port of Kobe ranks number one in Japan. Hence, the market behaviour of the port of Kobe is an attractive research topic. The transhipment cargo share through the port of Kobe was about 10 per cent of the total cargo in 2000. Transhipment cargo is considered as an import cargo volume for the purposes of this analysis.

11.2   Building a model of port behaviour

Export physical distribution function

The export physical distribution volume through the port of Kobe is hypothesized to be determined by five factors. These include three basic economic factors concerned with the port environment in Japan (E1–E3), reflecting the core of international trade theory. The two other factors (E4 and E5) represent the service level and the market power of the port of Kobe. Figure 11.1 depicts their simplified causal relationship.

The function of the export physical distribution volume (EPDV) for the port of Kobe can be written as follows:

$$\text{EPDV(Kobe)} = f\ (E1, E2, E3, E4 \text{ and } E5) \tag{1}$$

…where the three basic economic factors concerning the port environment and that are related to international trade theory are as follows:

- E1: Japanese direct investment in the following nine Asian countries and economies: Korea, Taiwan, Hong Kong, Singapore, China, Philippines, Indonesia, Thailand and Malaysia. In principle, the sign of the coefficient of E1 will be positive insofar as Japanese subsidiaries located in Asia continue to depend on the product procurement and sales activity of the Japan-based parent companies.
- E2: GDP (gross domestic product) of the nine Asian countries and economies. The coefficient should be positive.

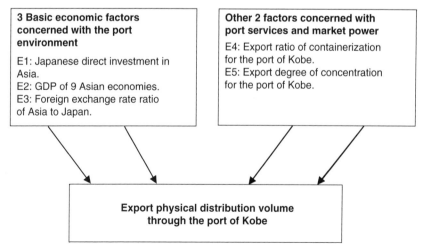

Figure 11.1   Export physical distribution framework found in the port of Kobe

- E3: Ratio of the foreign exchange rate of Japan to the nine Asian countries and economies. The coefficient should be positive.

The other two export factors concerned with port services and market power are as follows:

- E4: Ratio of export container cargo volume through the port of Kobe to its total cargo volume, which can be referred to as the 'Export Ratio of Containerization for the port of Kobe'. Its coefficient should be positive because it is assumed that a high level of port infrastructure will attract export container cargo.
- E5: Concentration ratio of export container cargo volume through the port of Kobe to total container volume in Japan, which can be referred to as the 'Export Degree of Concentration for the port of Kobe.' Its coefficient should have a positive sign.

The fourth and fifth factors (E4 and E5) are operational factors whose value depends upon the type of port management system utilized. In general, the greater the influence of the two operational factors on export container volume, the stronger the competitive position of the port of Kobe. In the long run, it can be assumed that Japanese ports will gradually lose their market share. Under such a process, a milder influence of the 'Export Degree of Concentration' would be more favourable.

Under the framework of this analysis, nine dummy variables representing the nine Asian nations and economies noted above are introduced in equation (1), so that:

$$EPDV(Kobe) = (1+DMi) \times f (E1, E2, E3, E4 \text{ and } E5) \qquad (2)$$

...where; DMi is a dummy variable representing the nine different nations and economies in Asia (i=1–8), including Taiwan (i=1), Hong Kong (i=2), Singapore (i=3), China (i=4), Philippines (i=5), Indonesia (i=6), Thailand (i=7) and Malaysia (i=8). Korea is the base nation in the estimation process.

Next, equation (2) is specified in the exponential form utilizing a log-log model. The coefficients of all the determinant factors (E1–E5) will have a positive sign.

The export distribution volume through the port of Osaka can be determined in principle in the same way as for the port of Kobe. Thus, the export physical distribution function for the port of Osaka can be written as follows:

$$EPDV(Osaka) = (1+DMi) \times f (E1, E2, E3, EO4 \text{ and } E5) \qquad (3)$$

...where E1, E2, E3, E5 and DMi are the same determinant factors as in the case of the port of Kobe and EO4 is the ratio of export container cargo volume in the port of Osaka to its total cargo volume, which can be referred to as the 'Export Ratio of Containerization for the port of Osaka'.

Utilizing the same specification method as equation (2), it follows that the coefficients of E1, E2, E3 and EO4 > 0 and that the coefficient of E5 < 0.

Import physical distribution function

The import physical distribution function can be developed utilizing the same logic as outlined above. Four basic economic factors concerned with the port environment, as well as two other factors concerned with port service and market power, determine import physical distribution volumes through the port of Kobe.

The framework for import physical distribution behaviour at the port of Kobe is described in Figure 11.2.

Hence, the import physical distribution volume (IPDV) through the port of Kobe can be written as:

$$\text{IPDV (Kobe)} = (1+DMi) \times f (M1, M2, M3, M4, M5 \text{ and } M6) \tag{4}$$

...where M1–M4 represent the four basic economic factors and M5 and M6 represent port service and market power respectively. The operations of the four basic economic import factors concerned with the port environment can be explained by the principles underpinning trade theory:

- M1 is Japanese direct investment in the nine Asian countries and economies under study. The sign of the coefficient will be positive

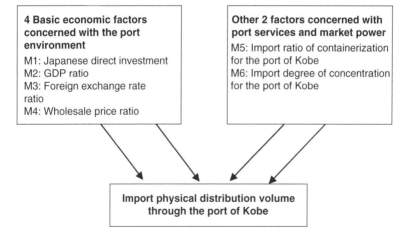

Figure 11.2 Import physical distribution framework for the port of Kobe

assuming rationally horizontal division of labour between Japan and Asia.

- M2 is the GDP ratio of the nine Asian countries and economies in relation to Japan. The sign of the coefficient will be negative.
- M3 is the foreign exchange rate ratio of Japan in relation to the nine Asian countries and economies. This coefficient will have a negative sign.
- M4 is the wholesale price ratio of the nine Asian countries and economies in relation to Japan. This coefficient will also have a negative sign.

The following two factors are concerned with port service and market power:

- M5: Import ratio of containerization for the port of Kobe.
- M6: Import degree of concentration for the port of Kobe.

In general, the sign of the coefficients will be positive in the case of M1, M5 and M6 and negative in the case of M2, M3 and M4.

In the case of the import function of the port of Osaka, the sign of MO6 (the import degree of concentration for the port of Osaka) will be negative, whereas the other determinant factors will have the same sign as in the case of the port of Kobe. Thus, it follows that:

$$IPDV\ (Osaka) = (1+DMi) \times f\ (\ M1,\ M2,\ M3,\ M4,\ M5\ and\ MO6) \qquad (5)$$

Utilizing the same specification as equation (4), it also follows that the coefficients of M1 and M5 > 0 and that the coefficients of M2, M3, M4 and MO6 < 0.

11.3 Seaborne trade patterns for the port of Kobe in relation to Asian physical distribution

This section aims to identify the characteristics of port functions by measuring the elasticity of export and import container cargo flows between the port of Kobe and the nine Asian countries and economies in relation to the determinant factors that generate the cargo flows (on a tonnage basis). In addition, this section is concerned with the export behaviour of the port of Osaka. The findings are broken down by country, as well as by exports and imports. The survey findings are likely to be useful in terms of identifying the defining characteristics of port management systems.

Characteristics of export behaviour of the ports of Kobe and Osaka

Export-related variables selected as factors causing changes in volumes of cargo through the port of Kobe include five determinant factors (E1–E5) as shown in Figure 11.1. On the other hand, import-related variables include six determinant factors as per Figure 11.2. The percentage point change for each factor was determined by measuring the elasticity of the change in total seaborne trade in relation to each of the variables under study. It is calculated by using econometric methods.

This study identified the following characteristics of export flows through the port of Kobe, which are summarized in Table 11.3. Initially, on the export side, gross domestic product (GDP) has the largest influence on export container volume flowing from Kobe to the nine Asian countries and economies under study. Thus, GDP is the determinant variable with the highest elasticity among the three main economic factors. The elasticity value of 1.045 indicates that a percentage point increase in the GDP growth rate in an economy results in a similar increase in seaborne trade for that economy. China and Thailand, however, are exceptions. The elasticity ratio for China is 1.336, while that for Thailand is below 1.0, at 0.779. This demonstrates that the Chinese economy was the driving force behind the overall Asian economy during the period under study.

A second point to emerge from the analytical results is that the elasticity of both direct investment and exchange rates are generally lower than the elasticity of the GDP variable. However, some economies have an elasticity above 1.0. A notable example is Thailand, which has by far the highest exchange rate elasticity, valued at 1.951. This is in accordance with the high fluctuation of its foreign exchange rate ratio. The figure is ten times larger than the normal level of elasticity coefficient of 0.196, which can be found in export trade with Korea, Hong Kong, China, Philippines and Indonesia. This suggests that the increased movement of export goods from the port of Kobe to Thailand is mainly the result of the rise in the baht-yen exchange rate ratio. The Thailand baht was extraordinarily overvalued in relation to the Japanese yen between 1986 and 1995.

It should be noted that the containerization ratio for the port of Kobe has the desirable effect of increasing export physical distribution volumes to all of the Asian economies, excluding Taiwan. This is especially true in the case of Singapore, with its elasticity coefficient of 2.122. Therefore, the port of Kobe's infrastructure is well suited to the technological progress apparent in container shipping.

Finally, in contrast, the port of Kobe's export share to Asian economies is decreasing under intense competition from other ports in Japan. The elasticity for the degree of export traffic concentration is 1.302 in all of the export trade routes to Asia. Thus, the port of Kobe loses 1.302 per cent of its export cargo volume for every one per cent decrease in export traffic concentration. Thus, it appears that the port of Kobe is unable to simply offer a higher qualitative level of infrastructure and expect to be market competitive. Why is the port of Kobe unable to achieve a high level of market performance? Does a barrier exist preventing service quality

Table 11.3 Export physical distribution function in the port of Kobe, 1986–95

Determinant factors	Coefficient	Korea	Taiwan	Hong Kong	Singapore	China	Philippines	Thailand	Indonesia	Malaysia
Japanese direct investment to Asia	Basic	0.291 (5.3)[a]	0.291 (5.3)[a]	0.291 (5.3)[a]	0.291 (5.3)[a]	0.291 (5.3)[a]	0.291 (5.3)[a]	0.291 (5.3)[a]	0.291 (5.3)[a]	0.291 (5.3)[a]
	Adjusted	—	—	-0.251 (-3.0)[a]	—	—	—	-0.289 (-2.1)[b]	0.176 (3.9)[a]	—
	Final	0.291	0.291	0.040	0.291	0.291	0.291	0.002	0.467	0.291
GDP of Asia	Basic	1.045 (9.7)[a]	1.045 (9.7)[a]	1.045 (9.7)[a]	1.045 (9.7)[a]	1.045 (9.7)[a]	1.045 (9.7)[a]	1.045 (9.7)[a]	1.045 (9.7)[a]	1.045 (9.7)[a]
	Adjusted	—	—	—	—	0.291 (3.0)[a]	0.070 (1.2)[e]	-0.266 (-1.9)[b]	—	—
	Final	1.045	1.045	1.045	1.045	1.336	1.115	0.77	1.045	1.045
Foreign exchange rate ratio	Basic	0.196 (1.8)[c]	0.196 (1.8)[c]	0.196 (1.8)[c]	0.196 (1.8)[c]	0.196 (1.8)[c]	0.196 (1.8)[c]	0.196 (1.8)[c]	0.196 (1.8)[c]	0.196 (1.8)[c]
	Adjusted	—	0.469 (-1.6)[d]	—	-0.369 (-1.8)[c]	—	—	1.755 (5.1)[a]	—	-0.044 (-2.7)[a]
	Final	0.196	0.273	0.196	-0.173	0.196	0.196	1.951	0.196	0.152
Ratio of containerization for Kobe	Basic	0.445 (4.5)[a]	0.445 (4.5)[a]	0.445 (4.5)[a]	0.445 (4.5)[a]	0.445 (4.5)[a]	0.445 (4.5)[a]	0.445 (4.5)[a]	0.445 (4.5)[a]	0.445 (4.5)[a]
	Adjusted	—	-0.460 (-4.5)[a]	—	1.677 (1.8)[c]	—	—	—	—	—
	Final	0.445	-0.05	0.445	2.122	0.445	0.445	0.445	0.445	0.445

Countries & economies

Table 11.3 Export physical distribution function in the port of Kobe, 1986–95 – *continued*

Determinant factors	Coefficient	Countries & economies								
		Korea	Taiwan	Hong Kong	Singapore	China	Philippines	Thailand	Indonesia	Malaysia
Degree of concentration for Kobe	Basic	1.302 (10.6)^a	1.302 (10.6)^a	1.302 (10.6)^a	1.302 (10.6)^a	1.302 (10.6)^a	1.302 (10.6)^a	1.302 (10.6)^a	1.302 (10.6)^a	1.302 (10.6)^a
	Adjusted	—	—	—	—	—	—	—	—	—
	Final	1.302	1.302	1.302	1.302	1.302	1.302	1.302	1.302	1.302
Constant		−7.74								

Statistical Result : RB2 = 0.915, SE = 0.21, N = 90

Note: The value in parentheses is the t-statistic. The superscripts a, b, c and d indicate the coefficients being significant at the 1%, 5%, 10%, 20% and 30% level respectively. RB2 denotes the coefficient of determination adjusted by the degrees of freedom, N the number of observations and SE the standard error of estimates.

Table 11.4 Elasticity of export physical distribution in the port of Osaka, 1986–95

Determinant Factors	Countries and economies								
	Korea	Taiwan	Hong Kong	Singapore	China	Philippines	Thailand	Indonesia	Malaysia
Japanese direct investment	0.318	0.318	0.318	0.318	0.318	1.192	-0.964	-1.471	0.318
GDP of Asia	0.727	0.727	0.727	0.727	0.727	-1.023	1.439	-1.517	0.727
Foreign exchange rate ratio	0.129	0.129	-0.569	-0.550	0.129	2.499	0.129	-8.465	0.129
Ratio of containerization for Osaka	0.060	0.060	0.060	0.060	0.060	-0.157	0.060	-0.916	0.606
Degree of concentration for Kobe	-2.22	-2.22	-2.22	-2.22	-2.22	-2.22	-2.22	-2.22	-3.12

Note: RB2=0.943, SE=0.05, N=90, Constant=1.616.

improvements by means of a large-scale increase in port infrastructure? These questions will be considered at a later stage in this chapter.

In the case of the port of Osaka, Table 11.4 indicates noticeably irrational market conduct in terms of direct investment in Thailand and Indonesia, as well as in the case of the GNP of the Philippines and Indonesia. In general, the port of Osaka has an unreasonable and strained economic relationship based on market principles with the four ASEAN countries, when compared with that of the port of Kobe. This suggests that export routes to the ASEAN countries from the port of Osaka appear to be irrationally established under competitive pressure from the port of Kobe and this has led to an accumulation of wasteful resources.

Secondly, the containerization ratio of the port of Osaka does not have as great an effect on its export physical distribution. The elasticity ratio is 0.06 – one-seventh the level of the port of Kobe's. The port of Osaka's infrastructure is not well suited to the technological progress of containerized shipping.

Thirdly, the port of Osaka suffers from intense competition from the port of Kobe. This is clearly exemplified by the –2.22 elasticity ratio representing the degree of concentration in the port of Kobe. However, in the reverse case, the port of Osaka will gain a 2.22 per cent increase in export container volume in response to a one per cent reduction in the market share of the port of Kobe.

Characteristics of import behaviour of the port of Kobe

The first noticeable characteristic on the import side is the negative elasticity of the GDP ratio (Asian GDP/Japanese GDP), except in the case of the Philippines and China (see Table 11.5). This indicates that imports into the port of Kobe will increase if the Japanese economy has high growth rates and is performing relatively well in comparison to other Asian economies. In general, this is attributable to the intra-industry trade relationship existing between Asia and Japan. However, in the case of Japanese imports from China, the reverse condition is true. The elasticity ratio of 0.347 indicates increasing imports, even though the Japanese economy is declining relative to that of China. This exemplifies the importance of, and interrelatedness between, the economies of China and Japan. This is also true in the case of the Philippines. Another noticeable result arises in the case of imports from Singapore. Imports from Singapore are extremely elastic at –2.956, indicating the strong influence of the Japanese economy.

Secondly, the coefficient of overseas direct investments by Japan has a positive effect on shipping activity. This is most likely due to a horizontal division of labour between Japan and Asia. This factor plays an important role as an index of the import logistics, except in the case of the Philippines. The intensity of this function varies from one economy to another, ranging between 0.163 and 0.451.

Thirdly, exchange rates show an anticipated negative value. The elasticity of exchange rates in relation to import volumes, in general, approaches the unit value of 1. The exceptions being the Philippines and Indonesia, where imports increase substantially due to the relationship between the strong yen and weak Asian currencies.

Fourthly, it should be noted that the containerization ratio for the port of Kobe has a neutral effect on the import physical distribution volume for all of the Asian countries and economies under study, excluding China. Thus, the containerization of the port of Kobe is not contributing to import logistics or import physical distribution volumes. While the service level of the port of Kobe is governing export logistics, it is having a neutral impact upon import logistics. The only exception is the case of China, for which it acts as a transhipment port.

Finally, if the degree of concentration of the containerized import cargo in the port of Kobe declines by one per cent, the actual amount of cargo from Asia decreases by 1.218 per cent. As already noted, the elasticity of the export degree of concentration in the port of Kobe was 1.302 for all of the nine Asian countries and economies being studied. This indicates that the port of Kobe's competitive position is deteriorating in terms of both imports and exports and that, therefore, a comprehensive restructuring of port management is necessary.

11.4   The strategic adaptation of port management

Core physical distribution behaviour in the port of Kobe

The estimation results in Tables 11.3 and 11.4 indicate that export physical distribution volumes can be determined by factors concerned with the economic environment as well as factors concerned with the quality of port service and market power. The results indicate that the market power of a port is affected not only by the economic development of its surrounding economic hinterland, but also by the port's own efforts to improve the quality of its services.

In contrast to Tables 11.3 and 11.4, Table 11.5 indicates that import physical distribution volumes cannot be determined by the quality of port service, although the other five factors influencing import physical distribution volumes do, indeed, have an effect on a shipper's logistics strategy. In the case of the port of Osaka and other ports in Japan, the same condition holds. Hence, in order to increase import distribution volumes at specific ports, it is important that a revolutionary industrial reorganization of the economic region surrounding the port works in combination with general economic development in Asia.

Table 11.5   Elasticity of import physical distribution in the port of Kobe, 1986–95

Determinant factors	Countries and economies								
	Korea	Taiwan	Hong Kong	Singapore	China	Philippines	Thailand	Indonesia	Malaysia
Japanese direct investment	0.451	0.451	0.261	0.451	0.163	-0.070	0.152	0.451	0.229
GDP ratio	-0.735	-0.735	-0.735	-2.956	0.348	0.079	-0.735	-0.375	-0.735
Foreign exchange rate ratio	-0.699	-0.699	-0.699	-0.699	-0.699	-1.443	-0.699	-2.085	-0.699
Wholesale price ratio	-0.616	-0.616	-0.616	-0.616	-0.616	-0.616	-0.616	-2.244	-0.616
Rate of Containerization for Kobe	0	0	0	0	2.375	0	0	0	0
Degree of concentration for Kobe	1.218	1.218	1.218	1.218	1.218	1.218	1.218	1.218	1.218

Note: RB2=0.943, SE=0.05, N=90, Constant=1.218.

Evaluation of current competitive strategy between the port of Kobe and the port of Osaka

In terms of export physical distribution volumes, the results of this study indicate that an aggressive strategy increasing the quality of port service can be evaluated as rational behaviour by Japanese ports. However, the current large-scale oversupply of infrastructure stems from the tremendously scattered investment policy pursued in the 1990s. Too much is as bad as too little.

The port of Kobe, which now boasts the highest levels of export container cargo in Japan, suffers from intense competition from the port of Osaka and other adjacent ports in western Japan. Thus, it will be generating negative economies of scale.

It would be more appropriate to combine the ports of Kobe and Osaka. The policy of developing export ports so as to avoid irrational behaviour entails grading each Japanese port in terms of the quality of services offered.

In terms of import physical distribution volumes, it is possible to conclude that the larger the economic strength of the region surrounding the port, the stronger the competitive power of the port under consideration. This suggests that the market power of the ports of Kobe and Osaka is underestimated under the current separate management systems. From this point of view, a merger or alliance of the two neighbouring ports will lead to improved market performance in comparison to the current competitive environment.

11.5   Evaluation of hypothetical ('virtual') type of merged management

The estimated results of five selected types of port management systems are compared in Table 11.6. These include the three hypothetical ('virtual') types of merged management systems (cases [1]–[3]) and the two current types of independent management systems (cases [4] and [5]). Korea is selected as the representative country of export movement from Japan to other Asian countries and economies.

Based on the estimated results in Table 11.6, it is possible to make the following three points:

1. The current type of management system as utilized by the port of Osaka (case [5]) results in a particularly low export ratio of containerization. This indicates that export shippers do not evaluate the service strategy of the port. As mentioned previously, this forces the port of Osaka to adopt an irrational strategy, one which is unsuitable for exports to the ASEAN countries. Under this scenario, intense competition with the port of Kobe generates an extensive waste of resources. Additionally, in case [3],

Table 11.6   Comparison of the elasticity coefficient under the five selected cases of export physical distribution management

Determinant Factor	Types of management				
	Case [1]	Case [2]	Case [3]	Case [4]	Case [5]
Investment of Japan	0.293 (6.32)[a]	0.245 (5.10)[a]	0.316 (6.37)[a]	0.219 (5.33)[a]	0.318 (2.68)[a]
GDP of Asian countries & economies	0.858 (11.6)[a]	1.090 (9.34)[a]	0.446 (8.21)[a]	1.045 (9.68)[a]	0.727 (6.21)[a]
Exchange rate ratio	0.387 (7.26)[a]	0.220 (3.53)[a]	0.177 (6.32)[a]	0.196 (1.76)[c]	0.129 (1.78)[c]
Ratio of container-ization	0.541 (3.63)[a]	0.150 (2.31)[b]	0.064 (3.63)[a]	0.445 (4.51)[a]	0.060 (1.67)[c]
Degree of concentration	0.442 (3.72)[a]	0.691 (5.92)[a]	0.414 (3.36)[a]	1.302 (10.6)[a]	−2.217 (−6.44)[a]

Note: Case [1]: $RB2=0.962$, $SE=0.19$, $N=90$, Constant $=-2.435$.
Case [2]: $RB2=0.950$, $SE=0.22$, $N=90$, Constant $=-5.523$.
Case [3]: $RB2=0.940$, $SE=0.24$, $N=90$, Constant $= 0.5167$.
Case [4] and Case [5]: Cf. Table 11.3, and Table 11.4.

where the port of Osaka manages the whole of the Osaka Bay Area, including the port of Kobe, the effect of this service strategy remains at the same low level as in case [5]. Therefore, the management type system of the port of Osaka is not appropriate for improving a shipper's evaluation of port service.

2. A second type of merged management system is case [2] where the port of Kobe has relatively dominant management power in the Osaka Bay Area. However, with the port of Kobe as the leader in a merged management system, the export rate of containerization is two-thirds less effective than in case [4]. This is similar to the management defect in case [3]. On the other hand, the lower elasticity ratio of the export degree of concentration indicates a smaller reduction in absolute export volumes from the Osaka Bay Area to other Asian countries in the event of increased competition from other ports located in western Japan. The lower level of elasticity in case [2] is an improvement upon the results generated by the current management type of case [4]. Thus, the port of Kobe benefits from this type of merged management system. However, from the shippers' viewpoint, it does not compensate for a lower reputation.

3. Case [1] represents the perfectly merged management system hypothetically ('virtually') adopted by the two ports. Under this system, a high

level of market performance is expected. In fact, Table 11.6 indicates that the elasticity of the export ratio of containerization attains the highest level of all five cases. In addition, the export degree of concentration functions with a mild level of elasticity, a desirable effect in a period of decline. In addition, the three determinant factors representing the economic environment of the port play a crucial role. If the three kinds of elasticity ratios representing the economic environment are summed, the total value is 1.538, which ranks second behind case [2] with an elasticity of 1.555. This difference is minimal compared with the 0.903 total value in case [3], 1.46 in case [4] and 1.174 in case [5]. From a different perspective, if the five-point estimation method in evaluating the three kinds of environmental factors is adopted, the point total of case [1] is eleven. This is equal to the level of case [2] and greater than the total in the other three cases. In the five-point estimation method, points are assigned based on the order rank of elasticity (from one point for the lowest elasticity up to five points for the highest).

In general, it appears that case [1] is equally as effective as case [2]. However, it should be noted that in case [1] a trade-off relationship does not exist with respect to the two operational strategic factors (the rate of containerization and the degree of concentration). To sum up, under an appropriate economic environment, the perfectly merged management system creates an upward shift in the production frontier curve which is composed of axes representing the two possible operational strategic factors. Thus, the potential exists for further development of the Osaka Bay Area through a change in management systems (see Figure 11.3). In case [1]

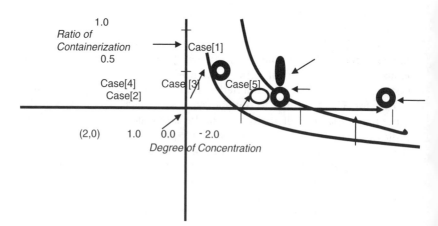

Figure 11.3 Evaluation of the selective type of management on the isoquants of the different levels of the production frontier

the port of Kobe and the port of Osaka are able to create an ideal union through the utilization of a merged management system.

11.6 Conclusions

A similar upward shift in the production frontier as noted above is not noticeably recognizable in the import physical distribution function. However, the perfectly merged management system causes the import ratio of containerization to operate with little variance from zero effects (the estimated results are omitted from this chapter). Therefore, the perfectly merged management system has no negative impact on the import distribution side.

However, a very steep barrier exists that prevents the initiation of this ideal policy. This is because in both cities the ports are managed under separate local self-governing bodies. The city of Kobe and city of Osaka both utilize local taxes to develop port management and increase infrastructure in order to compete with each other. Is it possible to merge only the port authorities of the two cities? The answer, at present, is no. Under the new governance concept it is necessary to enlarge and merge the local neighbouring autonomies in order to be able to utilize the more effective merged management system.[1]

In order to survive and compete under the present management system, it is particularly urgent for ports to improve service quality in order to respond to the needs of shippers' logistics strategies. It is also important to develop the economic hinterland surrounding the ports in addition to the Asian economy as a whole. If global shippers positively evaluate such revolutionary changes in the field of logistics, the ports will also be able to operate not only as a logistics service provider, but also as the aggressive system operator in the Asia-Pacific region.

Note

1 Discussions on this matter in Japan have been prominent following the IAME 2001 Conference in Hong Kong and then partially realized in 2003 as the Super Core Ports Project, which introduces the angle of the Osaka Bay area as the unified object of national port policy. We can recognise some progress in national policy, which should finally be combined with the ideal type of management mentioned above. Whether this will be possible or not, depends upon the joint efforts of the port of Kobe and the port of Osaka.

References

Miyashita, K., Tamura, M. and Tokutsu, I. (1999) Kitakyushu City: Functional Structure of the Physical Distribution Hub in the Asian Logistics Network, *East Asian Economic Perspectives*, 10, March, 38–120.

Miyashita, K. (1999a) *The Impact of the Asian Currency Crisis on Japanese Maritime Transport and its Environment*, OECD/DNME Workshop on Maritime Transport Policies, CNM/EMEF/MTC (99)1, October.

Miyashita, K. (1999b) *Strategy of Japanese Port Logistics Development on the Area of Kobe*, Proceedings of the International Logistics Symposium, Dongseo University, December.

Miyashita, K. (1999c) *Investigation of the Effects of Recent Environmental Changes on Shipping in Asia*, Tokyo: Ship & Ocean Foundation.

Ports and Harbour Bureau of Kobe City (1985–96) *Statistical Annual Overview of the port of Kobe* (Kobe-ko Taikan, in Japanese).

Ports and Harbour Bureau of Osaka City (1985–96) *Annual Statistical Report of the port of Osaka* (Osaka-ko Kosei, in Japanese).

12

Port Financing and Pricing in the EU: Theory, Politics and Reality[1]

H.E. Haralambides, A. Verbeke, E. Musso and M. Benacchio

12.1   Introduction

Increasing attention is being paid at the EU level to the desirability and scope of establishing a more harmonized European seaport financing and pricing strategy. A large-scale, pan-European research study for the European Commission (DG Transport and Energy), known by the acronym 'ATENCO' (Analysis of the main Trans-European Network ports' COst structures),[2] has been developed, with the main goal of providing input for an in-depth reflection at the European level on the design of a strategy to achieve efficient pricing and the possible impacts of a cost recovery approach to the functioning of ports.

Two recent policy orientations, at the European level, were crucial as starting points of this research study:

- The extension of the trans-European networks (TEN) to include, inter alia, seaports (see, e.g., the Proposal for a European Parliament and Council Decision No. 1692/96 EC as regards seaports, inland ports and intermodal terminals as well as project No. 8 in Annex III; COM (97), 681 final, 10 December 1997). The inclusion of these interconnection points is critical to the functioning of intermodal transport within a multimodal infrastructure network. The TEN will increase the options in terms of alternative door-to-door intermodal logistics chains available to transport organizers and users. In the more competitive environment faced by alternative logistics chains, distortions of trade flows between member states resulting from different systems of financing and charging for port (related) infrastructure and services could become or appear more important.
- Second, the Commission's Green Paper on Seaports and Maritime Infrastructure set out the broader context of Community port policy, with a focus on the issue of state aid and infrastructure charging (see: Green Paper on seaports and maritime infrastructure, COM (97) 678 final,

10 December 1997). Here, the main question was whether and how an efficient pricing system, leading to cost recovery, could be implemented in practice in the port sector, taking into account a variety of relevant objectives and constraints including higher market-based efficiency; increased cohesion; distributive goals; the development of short sea shipping; the improvement of safety and environmental performance, etc. Other, more recent policy documents at the European level have also addressed this issue; see, e.g. the Final Report (by the High Level Group on transport infrastructure and charging) concerning options for charging users directly for transport infrastructure operating costs.

At the time of writing, the Commission has developed its 'port package' (European Commission, 2001a and 2001b). Although the full implications of those two most important documents have yet to be fathomed, nevertheless it could be said at this point that the EC takes a fresh look at two recurring issues (among others):

- The need for greater transparency in the efficient allocation (leases/concessions) of port land to service providers on an equal opportunity basis and in a way by which leases reflect better the opportunity cost of port investments.
- The no longer indiscriminate treatment of port infrastructure investments as 'public investment'. Particularly with regard to the latter, although the Commission continues to remain neutral on the public or private ownership status of a port, and it does not dispute in any way the fact that public investments are the prerogative of member states, it nevertheless attempts to have a say in whether a certain investment, that in theory is open to all users indiscriminately but in practice it is intended for a few or even one user, could, in the spirit of the Treaty, be considered as 'public investment'.

Given the European policy directions described above, which could both be strengthened by the pan-European implementation of a coherent framework regarding port financing and pricing, the question arises of whether the adoption of any financing or pricing system or set of pricing principles at the European level would be a valid policy option.

12.2   Implications of the academic literature on the adoption of a European port pricing strategy

The main conclusion of a comprehensive academic literature review on port pricing (undertaken in the context of the ATENCO project) was that pricing in ports can and should be based on costs (Button, 1979). The determination of which costs should be reflected in prices largely depends on the type of port organization (Voorhamme and Winkelmans, 1980; Pollock, 1980; Suykens and Van de Voorde, 1998; Van Niekerk, 1996). Prices in service or comprehensive ports reflect a multitude of different costs – many of them joint costs, difficult to allocate in a way that is not largely arbitrary – compared to prices in landlord ports where more clear lines of responsibility and accountability exist (Thomas, 1978; Jansson and Shneerson, 1982; Verhoeff, 1981; Dowd and Fleming, 1994).

From a theoretical perspective, and assuming that a number of conditions are fulfilled, long-run marginal costs represent the most appropriate basis for efficient pricing (Walters, 1975; Bennathan and Walters, 1979). Alternatively, more sophisticated pricing arrangements, such as Ramsey pricing or two-part tariffs, have also been considered appropriate in the context of cost recovery in seaports (Ramsey, 1927; UNCTAD, 1975; Bennathan and Walters, 1979; Zachcial and Hautau, 1998). In practice, and in the absence of 'measurable' marginal costs, approaches based on average costs also appear to perform reasonably well in approximating marginal costs (Robinson, 1991), despite the problem of joint cost allocation and the inadequacy of port accounting systems (Gardner, 1977). In fact, differences between average cost pricing and marginal cost pricing are sometimes difficult to identify in practice (Talley, 1994; Coppejans and Bergantino, 1998) and as a consequence the discussion on the choice of an 'optimal' cost basis, although of academic interest, may in reality be of lesser practical significance.

Irrespective of the cost basis chosen, the principle that prices should accurately reflect (not to say recover) social opportunity costs[3] is crucial. Some of the things needed to achieve this are the collection of high-quality cost information by ports (UNCTAD, 1976; Thomas, 1978); greater transparency in port accounts and in the financial flows between the port and its institutional master; harmonization of accounting systems; and the adoption of a common glossary. Whether these data are subsequently interpreted as average cost data or as some approximation of marginal costs is a matter of little practical relevance, particularly in view of the fact that MC pricing is neither a necessary nor a sufficient condition for welfare maximization, as long as other related sectors (e.g. road/rail transport), impacting the port, are not priced similarly. In this particular context, a voice that is often loudly raised, by both the Commission (recently) and the port industry, argues that MC pricing in ports will only make port services 'unilaterally' more expensive, thus penalizing the Union's efforts to check road traffic and promote short sea shipping. A most valid argument indeed.

The policy implications of the above analysis are depicted in Figure 12.1 Here, the horizontal axis measures the (normative) requirement to apply marginal cost pricing as a precondition for welfare maximization: this requirement can be viewed as weak or strong. The vertical axis reflects the requirement to collect, process and effectively interpret complex, high-quality cost

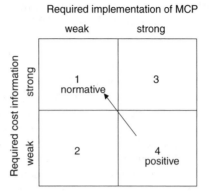

Figure 12.1   Preconditions for efficient pricing

information as a precondition to achieving cost recovery: here again, this requirement can be either weak or strong.

A careful analysis of European policy documents on efficient port pricing, especially the Green Paper on Ports and Maritime Infrastructure, largely supports the adoption of a quadrant four perspective. The implementation of marginal cost pricing is indeed viewed as a key requirement to achieve efficient pricing in ports, albeit subject to a number of exogenous policy considerations, mainly of a distributive nature. Less attention is paid to the administrative problem of collecting, processing and interpreting complex, high-quality cost information, most likely because this is viewed as an operational issue rather than a strategic policy one.

The above analysis suggests, however, that a quadrant one perspective should be adopted. As has already been mentioned, there is no strong requirement to apply marginal cost pricing as a precondition to achieve efficient pricing in seaports. However, any policy proposal at the European level aimed to foster the systematic application of specific cost recovery rules, should – and this is a strong requirement – focus on the condition of high-quality cost information availability, which obviously implies, for example, uniform definitions of cost and income components as well as balance sheet items, and transparent accounting rules.

In addition to the principle of cost recovery, the increasing transformation of ports (at least of competing regional container terminals) from public to private enterprises raises the issue of the desirability and fairness of pricing methods focusing on the 'user' rather than the 'general taxpayer' (EU COM (97) 678 final). Despite the continuing importance of the joint cost allocation problem here too (particularly here too one might say), the allocative and income distribution effects of such a switch in direction are obvious: investments are recovered, and port revenues generated, from the user of a (private) facility, who will have to somehow pass these costs on to the final consumer. In all likelihood the latter will have to pay higher prices for the goods he consumes but, at least in efficient markets, he is compensated by correspondingly paying less taxes (for infrastructure investments). Obviously, such issues are highly complex and have yet to be fully researched.

12.3 The practice of port pricing in European ports

The analysis of present pricing strategies in European ports, examined as part of the ATENCO study, demonstrated the substantial diversity prevailing among European Union ports with regard to their financing and charging practices. This diversity is deeply rooted in the various judicial and cultural traditions as well as in the divergent port management styles, related responsibilities and degree of autonomy. These observations are fully consistent with the conclusions of previous fact-finding reports of the European Seaports Organization (ESPO) and many academic studies (Heggie, 1974; Pollock, 1980; Suykens, 1986).

This observed diversity makes the requirement of uniform methods of cost recovery and related pricing a very complex issue, and it suggests a gradual, step-by-step approach. Undoubtedly, such an approach should respect, at least in the short run, national perceptions on the appropriate role of public investment, still the prerogative of member states.

However, considerable progress could be made through efforts aimed at harmonizing definitions and classifications of port infrastructure. Current classifications (e.g. investments inside or outside the port area) often lack an economic rationale and are instead based on technical or geographical elements (e.g. to determine whether investment costs should be allocated directly to users or to society at large).

Giving ports greater financial autonomy could contribute substantially towards achieving cost recovery through whatever pricing might be deemed appropriate by the ports themselves. However, obtaining such autonomy usually depends upon a political decision making process. Furthermore, in order to avoid the conventional drawbacks of monopolistic situations, such autonomy should be accompanied by efforts to ensure adequate intra-port competition, either at the operational level or through rent/lease policies.

Finally, as far as nautical services are concerned (pilotage, mooring and towage), MC pricing is still a pertinent issue albeit in a different perspective, this being the retention of economic rent for society, public service obligations (PSO) notwithstanding. As a result of the very nature of the services involved, the often limited size of the market (and thus sometimes the (im)possibility to increase competition) and port safety considerations, nautical services are often either carried out by public agencies or by private parties with exclusive rights. The potential for the abuse of a dominant position is thus real and this problem can be addressed either through encouraging competition or through more effective price control. Here, successful MC pricing is able to establish 'cost-relatedness' at the same time.

However, the real challenge of any policy attempt to foster some type of EU-wide cost recovery discipline is to come to grips with the dependency of port financing and pricing routines on both (sub) national institutional settings, external to ports, and (sub) national political decision making which, *ad passim*, may use ports as a lever to achieve broader policy goals. The former parameter includes elements related to the judicial, cultural, organizational and managerial heritage of a seaport, including traditions related to the allocation of responsibility and authority to the various actors functioning in the seaport system. The latter parameter refers to broader policy goals, often pursued by public policy makers, such as growth pole effects, employment, and regional value added, distributive equity, etc. In the highly sensitive European port environment, such parameters and pursuant policies are strong enough to negate any radical centralized policy initiatives, however appropriate these may be in an increasingly integrated Europe.

Both parameters imply that, in order to make some headway in formulating institutional change, substantial attention should be devoted to historical trajectories and path dependencies associated with specific (sub) national port financing and pricing routines. If not, the danger exists that substantial unintended policy effects might arise. If the key goal of a European policy initiative in the cost recovery area is the establishment of a 'level playing field' among competing European ports, it should be recognized that any assessment of potential improvements cannot be solely undertaken in terms of purely-market based considerations, in contrast to many other sectors where liberalization and market-based rules have been widely credited as being instrumental in the creation of better and best practices.

In the seaport sector, where several actors may be involved in the vertical port activities chain and the horizontal port activity clusters, the situation may be very different, precisely because of the potential of unintended policy impacts. For instance, a different financing and pricing discipline externally imposed on ports may disturb the effective horizontal and vertical linkages among the various port actors.

In spite of its normative emphasis on efficient pricing, the Green Paper did not fail to demonstrate a deep awareness of port 'dependences' and broad economic policy objectives. Unfortunately, the recognition of such complexities was not equally apparent in the approach of the so-called High-Level Group on Transport Infrastructure and Charging in its final report on Options for Charging Users Directly for Transport Infrastructure Operating Costs (EU, 1998).

Clearly, however, considerations such as those cited above do not stand up to the 'surgical' test of economic rationality from the point of view of an impartial and uninvolved onlooker, the more so when it becomes apparent that such arguments, apart from being often inadequately defended, are at times used as pretexts and conscientious impediments against making more headway in establishing a level playing field among competing ports. The challenge of any policy (and this is propounded in a forthcoming paper) is to reconcile, in a win–win situation, two things:

- The public with the private interest; and
- The pursuance of wider socio-economic objectives, with the nowadays paramount need to allocate (European) resources effectively.

12.4 A strategic framework for the analysis of financing and pricing practices in EU ports and the expected 'value added' of an EU-level approach in these matters

The objective of the ATENCO study was to explore the impacts of various pricing scenarios on the competitive position of seaports, with a view to identifying generally accepted pricing strategies that could lead, in the long run, to the efficient allocation of scarce resources and, consequently, to the establishment of a level playing field among competing European ports.

Hence, it appeared reasonable to attempt to define, from a conceptual perspective, only two extreme generic types of ports. First, the type that could relatively easily implement a charging framework aimed at full cost recovery (but perhaps one with a pricing system that would already be largely in line with such a goal). Secondly, the type that would, for a variety of reasons, encounter major difficulties in doing so.

Based on extensive and in-depth analyses of the functioning of several European ports, the authors identified five parameters critical in distinguishing between Generic Type I and Generic Type II ports.[4]

Ownership

If the public sector has an ownership stake in the port authority, this usually implies that some form of political input enters the prevailing decision-making process of the port, including decisions on investments, wage structures and thereby (even if only implicitly) also decisions on pricing. Even if the public sector involvement only aims to reduce or eliminate market failure, such considerations may make microeconomic decision making in the pricing area more complex. A higher number of public institutions as owners is likely to increase this problem even further.

Objectives

Irrespective of the ownership structure, the question arises whether the port's key goals are centred on microeconomic issues such as profitability and market share considerations or whether broader macroeconomic and societal goals such as regional economic development or the international competitiveness of domestic industries are being pursued. In the latter case, important effects can again be expected on the pricing decisions of the port.

Autonomy

Irrespective of the port's ownership structure and key objectives, autonomy reflects the extent to which the port can perform port related activities independently, without being unduly constrained by boundary conditions imposed by external actors (national or regional regulations, municipal constraints, etc.), beyond those applicable to conventional commercial undertakings. Again, the existence of external constraints may greatly influence pricing schemes and decisions.

Scope of activities

Here, the question arises as to what extent the port is responsible for simultaneously carrying out activities that can be run according to conventional commercial principles, and activities that either entail an important regulatory component (e.g., because they include health and safety considerations or aim to eliminate potential negative impacts of monopoly power) or are difficult to perform on a commercial basis for other reasons (e.g., overall port strategic planning and dredging work). It is precisely when a port engages in both types of activities simultaneously that pricing effects may occur in those activities that can (should), in principle, be commercially run. For example, cross-subsidization may then occur between the non-commercially-run and commercially-run activities.

Public support à fonds perdus

Finally, perhaps the most critical parameter that could potentially affect a port's pricing behaviour is whether public funds are allocated *à fonds perdus*, i.e., without any formal *ex post* performance requirements or systematic performance evaluations. The ATENCO study identified the sources of investments for several types of infrastructure as well as superstructure in various EU ports. It appeared that public investment (as well as other types of support) is still important and can be found in many EU countries. Obviously, the involvement of public funds, of whatever nature, does not necessarily imply that port users get free or low-priced access to these infrastructures and superstructures. It does imply, however, that the incentive to engage in some type of cost-based pricing becomes blunted to say the least.

Table 12.1 Generic type of port authority

Parameter	Type I	Type II
1. Ownership	Private	Public involvement
2. Objectives	Profitability/market share	Broader societal goals
3. Autonomy	Strong	Weak
4. Scope of activities	Mainly commercially run	Mixed (commercial and non-commercial)
5. Public resources allocation *à fonds perdus*	None/low	High

The two generic types of port authority are described in Table 12.1. Ownership, autonomy, scope of activities and the allocation of public resources *à fonds perdus* can be viewed as the key components of a (sub) national port system's institutional setting.

If, in practice, important deviations are observed from some form of cost-based pricing for infrastructure and superstructure, the question which needs to be considered is which of the five parameters or what configuration of parameters (closer to Type II) may be most relevant to explain specific pricing practices.

Perhaps more importantly, the above analysis suggests that if the introduction of a pan-European level, cost-based pricing framework were contemplated, the question needs to be asked which of the five generic port type parameters (ownership, objectives, autonomy, scope of activities, public resources allocated *à fonds perdus*) would need to be changed at the level of each port (and national port system) and what would be the implications for the port's (and national port system's) functioning. This would precisely allow the identification of likely unintended policy impacts.

12.5 Empirical evaluation of the validity of adopting a European port pricing strategy

In the ATENCO study, a survey questionnaire was developed, with the help of the European Commission services, that aimed to gather information both on present pricing principles and strategies, and also on the likely impact of introducing new pricing systems. In fact, two questionnaires were developed: the first to be completed by port authorities and the second by port users. The results of the survey are briefly discussed below.

Port authorities and port users in 13 European TEN-T ports were interviewed about their pricing practices as well as their economic performance. Their views were also asked on competition and a common framework for 'user pays' principles for the financing of port infrastructure.

These ports constitute a key sample of the largest ports in European countries. Not all ports were able to provide answers to all questions.

For many ports the reasons for failing to answer a specific question were related to a lack of knowledge or information within the port. The various European ports have very different accounting systems and practices for the registration of activities.

The ports included in the survey were the following: Aarhus, Denmark; Antwerp, Belgium; Barcelona, Spain; UK Port B, United Kingdom; Dublin, Ireland; UK Port A, United Kingdom; Genoa, Italy; Goltherburg, Sweden; Hamburg, Germany; Lisbon, Portugal; Piraeus, Greece; Rotterdam, The Netherlands; and Venezia, Italy.

The following five conclusions of the survey appear to be particularly important:

- All port authorities supported the adoption of overall full cost recovery within the port sector and considered it to be at least of some impor- tance. Five ports even considered full cost recovery to be of critical importance for individual activities. The majority of the ports supported the adoption of 'user pays' principles in ports. Surprisingly, most port authorities expected that the adoption of full cost recovery pricing would have little impact on pricing levels. It is believed here that, although in private ports such as those of the UK this may well be the case, this is far from the case in all others, and this conviction of many port managers can only be explained by their inability to grasp the full implications of long-run marginal costs.
- The port authorities did not consider the markets for liquid and dry bulk cargoes to be influenced by public support schemes. The markets for general cargo were considered by some ports to be influenced by such schemes, while most of the ports considered that the markets for containerized and Ro-Ro cargo were influenced by these schemes. Most of the ports believe that a more rigorous adherence to cost recovery would be beneficial to the port sector. Seven of them were even in favour of the uniform adoption of general pricing principles to the extent, however, that adherence to these principles would still allow flexibility and that hinterland transport pricing should be subject to similar principles.
- The port users were generally aware of some impact or distortion caused by public support schemes in European ports. The users considered the impact to be of limited relevance in relation to the prices charged by the port operators to the users and of some importance in relation to the overall port user costs.
- The interviewed users stated that the market for liquid bulk is inelastic in relation to port user costs, even for large variations in the price (up to +/– 50 per cent). Dry bulk cargoes were assessed to be inelastic for small changes and elastic for large changes (+/– 15 to 50 percent). General cargoes were assessed as elastic even for relatively small changes in port

user costs. The container market was considered to be inelastic for small changes and elastic for large changes. The Ro-Ro market was considered to be more inelastic to small changes than the container market, but elastic to large changes (+/– 15 to 50 per cent).[5]

- Port users disagreed about the influence of public support schemes in the fields of liquid and dry bulk, whereas they all shared the belief that the markets for general cargo, containers and Ro-Ro cargo were subject to some degree of influence from public support schemes. The users were not satisfied with the present port pricing policies and they believed that a more rigorous adherence to cost recovery principles would most likely be beneficial to the European ports. The adoption of specific cost recovery rules was viewed as likely to be beneficial. However, almost all users were opposed to uniform pricing that would be imposed by governments, if that were to ever be the case.

The above elements allow one to conclude that the generalized adoption of full cost recovery principles at the level of each port authority's entire set of activities under its control is viewed as desirable by all of the port authorities surveyed. A corresponding pricing principle could therefore be used as a starting point for discussions between the European Commission and the (sub)national agencies responsible for port policy. In addition, all port authorities and port users viewed clear and transparent linkages between costs and pricing as important and even necessary. Hence, policy efforts at the European level aimed at increasing this clarity and transparency will be welcomed in principle throughout the port sector.

The concept of 'flexibility' represents the preference of many ports and port users for new pricing principles aimed at eliminating present distortions of competition, where such distortions are non-trivial. Here, two key factors in achieving success will be the ability of the Commission to distinguish between the presence of such non-trivial distortions of competition on the one hand and the existence of widely diverging pricing practices among ports, not associated with such distortions, on the other hand. Equivalent treatment represents the simultaneous introduction in the hinterland modes of exactly the same pricing principles as in the port sector.

If the two conditions given above cannot be respected in practice, it may be desirable to shift European policy activities from designing a charging framework in the pricing area to:

- Defining acceptable public financing practices for port infrastructure, superstructure and services (*ex ante* prevention of distortions); and
- Rethinking how conventional competition rules (related, inter alia, to market access, the abuse of dominant positions, collusive behaviour, etc.) should be applied to the port sector (*ex post* sanctioning of distortions).

Table 12.2   Price elasticity for selected North Range container ports (10 per cent price increase; simulation results)

Port	Elasticity
Hamburg	3.1
Bremen Ports	4.4
Rotterdam	1.5
Antwerp	4.1
Le Havre	1.1

The survey described above was complemented with a quantitative simulation exercise building upon proprietary modelling tools of the Bremen Institute of Shipping and Logistics (ISL) and with a special focus on container flows in Europe. The analysis was based on ISL's model 'A simulation and forecasting model of world containers shipping including port hinterland traffic' (Bremen, 1997).

The simulation exercise analysed how different pricing schemes would affect traffic volumes in individual ports. Container traffic via the North Range ports was chosen as the core case because of the availability of the necessary data and the fact that ISL had already performed several simulation projects for container transport in Europe and therefore had developed a database with origin/destination points of European container flows as well as the related transport costs. The simulation led to three important conclusions:

- The price elasticities for container traffic diverge substantially among European ports, as illustrated in Table 12.2.
  Whether the absolute level of these elasticities is correct is a much less important issue than the observation of a very substantial divergence of the elasticities among the various ports. Hence, variation in prices, as a result of the adoption of alternative pricing systems, would, at least in the case of containers, lead to fundamentally different impacts on individual ports, even when engaging in similar price increases.
- The price elasticities appear – unsuprisingly – to vary considerably across cargo categories. More specifically, the price elasticities are in general much lower for liquid and dry bulk than they are for containers, general cargo and Ro-Ro. Given that government support schemes and distortions of competition are perceived as relatively unimportant for the former cargo categories but rather important for the latter, the introduction of cost-based pricing is likely to affect precisely those traffic categories that are most price elastic.
- If the introduction of new pricing principles were to focus on overall full cost recovery at the level of the individual port, those ports with a

substantial share of bulk cargo and high income from land rentals from industrial companies would be able to largely compensate any resulting price increases in the container, general cargo and Ro-Ro areas through cross-subsidization.

The above analysis implies that an across-the-board adherence to a specific pricing discipline may be expected to bring equality to the European port scene in the long run. In practice, however, the short term implications for the traffic market share and the income of individual ports will vary substantially, depending upon factors not entirely related to the magnitude of present government support levels or observed distortions to competition.

The empirical research was concluded with two sets of case studies on the impact of adopting the cost recovery approach in port financing and charging. The first set consisted of case studies of ports where cost recovery principles are already largely implemented. The second set included case studies of ports where the adoption of cost recovery principles is not viewed as critical.

The first set examined three ports where charges are determined on the basis of full cost recovery: two in the UK and one in the Republic of Ireland. The ports are UK Port A, UK Port B[6] and Dublin. The review of these case studies led to the following five conclusions.

Although, in principle, each port seeks full cost recovery at the level of both overall financial performance and the performance of specific profit niches, several other pricing principles are applied in practice, including pricing in view of competition, pricing according to 'what the traffic can bear', pricing as a function of capacity utilization, and so on. This is an important observation as it suggests that even ports seeking overall full cost recovery apply a variety of pricing principles simultaneously, in order to achieve managerial effectiveness at the micro-level.

Among the three cases included in the study, major differences in pricing strategy exist, as a result of each port's institutional heritage and managerial objectives. This suggests that even when the presence of various pricing principles is taken into account, the actual mix of pricing principles adopted in pricing strategy implementation may vary substantially, even when full cost recovery is pursued, as a result of managerial discretion.

In contrast to the widely held belief that UK and Irish ports engage in conventional full cost recovery, the study found that users in fact do not pay for past capital investments in terms of their replacement value.

The inclusion in prices of external costs caused by port activities appears to be a very complex issue. All three port authorities share the view that the costs they have incurred through their compliance with EU, national and international legislation on safety, health and environmental standards and their commitment to various related voluntary codes or practices have resulted, to some extent, in an internalization of the external costs that their business activities impose upon society and are thus reflected in their present charges.

There is a wide divergence of opinion, though, about the extent to which they consider present charges to reflect the internalization of all external costs. This divergence of opinion may be explained partly by subjective perceptions regarding the port authority's responsibility for the external costs imposed on society by port-related activities and the scope of activities to be included. One of the key issues here is whether or not port charges should reflect the external cost of congestion to which port-related traffic undoubtedly contributes and how this should be done to reflect the 'polluter pays' principle.

The case studies demonstrate that the hypothetical introduction of government financial support, similar to support mechanisms that exist in continental ports, would lead to very different effects: (a) on the various ports; (b) on the various types of port operations depending, inter alia, upon their traffic mix, cargo volumes and stevedoring costs; (c) on the various shipping companies, depending upon the number of calls made to UK and Irish ports, the size and type of the vessels involved, the share of total cargo loaded/unloaded in UK and Irish ports and the share of cargo handling costs in overall port user costs.

In any case, the hypothetical introduction of government support would be unlikely to greatly improve the various ports' competitive position and, perhaps more importantly, it would not alter their marketing strategy.

The case studies of ports practicing full cost recovery demonstrate the presence of a wide variety of pricing principles used in practice. The pricing strategies of these ports exhibit substantial managerial discretion that cannot be captured fully by textbook definitions of pricing. A best practice formula for pricing in the real world clearly does not exist, not even in those ports pursuing full cost recovery as a primary objective.

The second set of case studies (the ports that do not focus primarily on cost recovery) included the North Sea container ports, the Mediterranean transhipment ports and the linkages between the Iberian Peninsula and North-West Europe in the Ro-Ro area.

These case studies yield the three following conclusions:

- The case study on the North Sea container ports focussed on the likely impact of full cost recovery of maintenance dredging in the ports of Rotterdam, Antwerp, Hamburg and Bremen. Here, the ports of Rotterdam and Antwerp appeared to emerge as winners with an expected gain in traffic, whereas the German ports would lose traffic. A more general approach, which evaluated the impact of price increases in each of these four ports vis-à-vis the three others, showed very different effects in each case. For example, a price increase in Rotterdam would clearly primarily benefit Antwerp (and vice versa), whereas a price increase in one of the German ports would primarily benefit the second German port.

• The possibility of cost recovery in the Mediterranean transhipment hubs is a very complex issue for several reasons. First, the structure of trade in the region is changing very rapidly. The domination by shipping companies with a home base in the Mediterranean is being replaced by the entry of global carriers, especially wih regard to the Europe–Far East routes. As a result, shipping costs per unit are decreasing and more attention is being paid to scale economies and to calling at new container hubs rather than at those ports that have been historically significant in the region. Second, the new transhipment hubs are viewed as instrumental to the economic development of the less favoured regions in which they are located. As a result, the development of some of them (e.g. Gioia Tauro) has been assisted by Community funding. Third, although a long-term balance between demand for and supply of container handling capacity is expected, at present substantial overcapacity exists, which contributes to a high volatility of market shares of individual ports and port operators. The introduction of (full) cost recovery in the short run is viewed as non-feasible because strong rivalry puts tremendous pressure on prices.

• The case study of Ro-Ro services linking the Iberian peninsula with North-West Europe suggests that, in the long run, additional port costs, resulting from full cost-based pricing, are unlikely to have much impact on the viability of the Ro-Ro services. However, in the short run, when starting up new operations, port price increases could reduce the competitiveness of short sea shipping vis-à-vis road transport. In more general terms, the policy objective of full cost recovery in ports appears to be in conflict with the goal of promoting short sea shipping vis-à-vis road haulage and with improving the accessibility of peripheral areas.

In conclusion, the number of parameters critical to the assessment of the impact of cost-based pricing in European seaports is so large that only an in-depth analysis of all relevant case studies in terms of single traffic categories in individual ports can lead to a correct and comprehensive overview of the new pricing principles' effects. Given the practical difficulties associated with such a bottom-up approach, bounded rationality constraints suggest that it may be sensible to shift policy attention from emphasizing the importance of adopting uniform cost-based pricing principles (the so called charging framework) toward focussing on the more indirect incentives promoting cost-based thinking in ports (e.g. by more clearly defining what constitutes acceptable public support of port infrastructure and superstructure).

12.6   Conclusions

The issue of infrastructure pricing, of ports in particular, is highly complex. In Europe, the socioeconomic objectives often pursued by ports are so divergent that any uniform approach to pricing becomes meaningless and politically unfeasible. On the other hand, at least in a liberal economic environment such as that of the EU, pricing matters ought to be, ideally, left to the producers (ports) themselves.

The issue of port pricing – and the Commission's involvement in it – has not arisen out of academic curiosity but rather as a response to the need felt in the port industry itself for a self-discipline mechanism that, if consistently applied, would eventually lead to the recovery of port investments and to future investments that are largely demand-driven. This requirement has been the result of the recognition that, in today's intensified regional port competition and the increasingly tightened fiscal constraints of an integrated Europe, it is no longer acceptable to indiscriminately, and with no formal economic rationale, spend taxpayer money on port investments, often aimed at increasing market share at the expense of other ports, particularly those in neighbouring member states.

The ATENCO study has demonstrated that, however controversial the issue of port pricing itself may be, there is general consensus on the importance of cost recovery. And this marks an important development and step forward. Indeed, as long as this objective is respected, the specific pricing policy of the individual port becomes of secondary importance and only of interest in so far as crowding-out effects and efficient allocation of resources are concerned. But these, so far, are matters of national rather than European economic policy.

Once cost recovery is generally accepted as a guiding principle in port investment and pricing, the way forward is much simpler. It involves the compilation of better and more harmonized statistics on port costs, the adoption of standardized port accounting systems, greater transparency of port accounts and of financial flows between the port and its institutional master and, perhaps, a common glossary of terms. And these are objectives that are not so difficult to achieve.

Notes

1   This chapter was previously published as Haralambides et al. (2001). The editors are grateful to Palgrave Macmillan for granting permission to reproduce it herein.
2   The authors have been involved in various capacities in this project and one of the purposes of this chapter is dissemination, according to EU requirements. Notwithstanding this, views and opinions expressed herein are those of the authors only.

3   Defined here as the costs of the factors of production (exclusive of possible economic rent) required to produce the port service. This definition, particularly the word 'social' does not have to necessarily include external costs of production, something that has often been a cause of confusion.
4   A sixth relevant parameter – namely port networking – is not taken into account here, given that it may be largely driven by forces other than the port authority or the public agencies responsible for port policy.
5   Until further economic analysis is undertaken, such conclusions have to be viewed with considerable caution and an understanding of the fact that many port managers do not have formal training in economics. The notion of 'price= =elasticity' is not always easy to grasp in full, particularly the fact that the elasticity of demand for a good or service is not, in principle, a function of how large or small the change in price is. Perhaps the questionnaires could have been different, or clearer, on this point as well as on the point of long-run marginal costs.
6   We were informed by one of the paper's referees that the two UK ports did not wish their names to be revealed.

References

Bennathan, E. and Walters, A.A. (1979) *Port Pricing and Investment Policy for Developing Countries*, Oxford: Oxford University Press.

Button, K.J. (1979) The Economics of Port Pricing, *Maritime Policy and Management*, 6(3), 201–7.

Coppejans, L. and Bergantino, A. (1998) Economic Considerations With Respect To Port Pricing, Cardiff University Report.

Dowd, T.J. and Fleming, D. (1994) Port pricing, *Maritime Policy and Management*, 21(1), 29–35.

European Commission (1997) *Green Paper on Sea Ports and Maritime Infrastructure*, 678 final, 10 December, Brussels: European Commission.

European Commission (1998) *White Paper on Fair Payment for Infrastructure Use: A Phased Approach to a CommonTransport Infrastructure Charging Framework in EU*, 466 final, Brussels: European Commission.

European Commission (2001a) Communication from the Commission to the European Parliament and the Council: 'Reinforcing Quality Service in Sea Ports: A Key for European Transport'. Proposal for a Directive of the European Parliament and of the Council on Market Access to Port Services. COM(2001) 35 final, Brussels, 13 February.

European Commission (2001b) Commission Staff Working Document on Public Financing and Charging Practices in the Community Sea Port Sector. SEC (2001), Brussels, 14 February.

Gardner, B. (1977) Port Pricing – an Alternative Approach, *Transport of Steel Exports – an Investigation into the Scope for Rationalization*, vol. I, chapter 4.2, Cardiff: University of Cardiff, Department of Maritime Studies, Transport Research Unit.

Haralambides, H.E., Verbeke, E., Musso, E. and Benacchio, M. (2001) Port Financing and Pricing in the European Union: Theory, Politics and Reality, *International Journal of Maritime Economics*, 4(3), 368–86.

Heggie, I.G. (1974) Charging for Port Facilities, *Journal of Transport Economics and Policy*, 8, 3–25.

Jansson, J.O. and Shneerson, D. (1982) *Port Economics*, Cambridge, MA: MIT Press.

Pollock, E.E. (1980) Port Tariff Policy in Europe – A British View, *Tijdschrift voor de vervoerswetenshap*, 199–204.

Ramsey, F. (1927) A Contribution to the Theory of Taxation, *Economic Journal*, 37, 47–61.

Robinson, R. (1991) Pricing Port Services in Australia – the Issues, paper presented to New Thinking on Port Pricing, Executive Development Programme, University of Wollongong, Centre for Transport Analysis.

Suykens, F. (1986) Ports Should be Efficient (Even if This Means That Some of Them are Subsidized), *Maritime Policy and Management*, 13(2), 105–26.

Suykens, F. and Van de Voorde, E. (1998) A Quarter of a Century of Port Management in Europe: Objectives and Tools, *Maritime Policy and Management*, 25(3), 251–61.

Talley, W.K. (1994) Port Pricing: a Cost Axiomatic Approach, *Maritime Policy and Management*, 21(1), 61–76.

Thomas, B.J. (1978) Port Charging Practices, *Maritime Policy and Management*, 5(2), 117–132.

UNCTAD (1975) *Port Pricing*, TD/B/C.4/110/Rev.1, New York: United Nations.

UNCTAD (1976) *Manual on Port Management – Part Four (Modern Management techniques): Port Pricing* , New York: United Nations.

Van Niekerk, H.C. (1996) Efficient Pricing for Public Ports, IAME International Conference, Vancouver.

Verhoeff, J.M. (1981) Seaport Competition: Some Fundamental and Political Aspects, *Maritime Policy and Management*, 8(1), 49–60

Voorhamme, R. and Winkelmans,W. (1980) Port Tariff Making in 10 EEC Sea Ports, *Tijdschrift voor de vervoerswetenshap*, 253–72.

Walters, A.A. (1975) Marginal Cost Pricing in Ports, *The Logistics and Transportation Review*, 11(4), 299–308.

Zachcial, M. and Hantau, M.U. (1998) *Relevant PricingTechniques Applied to Port and Maritime Infra- and Superstructure*, report for Institute for Shipping Economics and Logistics, Bremen.

13

A Hierarchical Taxonomy of Container Ports in China and the Implications for their Development[1]

Kevin Cullinane, Sharon Cullinane and Teng-Fei Wang

13.1 Introduction

Prompted by a phenomenal growth in trade over the last decade, China's container ports have been increasing in both number and importance. Within the Asian market sector, container ports in the Chinese mainland now threaten to undermine the dominance of both Hong Kong and Singapore. This chapter describes the development of China's container ports to this point in time and, by applying a classification system based on a hierarchy of ports, seeks to deduce likely scenarios for the sector's future development.

The development of Asian hub/feeder networks has been divided into three phases by Robinson (1998). These phases are differentiated by the degree of complexity in the structure of the port hierarchy within the region. Specifically, from 1970 to the mid-1980s, there were only a few, very important, hub ports which existed within a single hierarchical level and which dominated the trade in a liner service network configuration that stretched from Singapore to Japan.

With the transition into the next phase, identified by Robinson (1998) as taking place over the period from the mid-1980s to the mid-1990s, a second tier of regionally significant feeder ports came into being and added a new dimension to Asia's liner service networks. From the mid-1990s to 2000, the third phase was characterized by the expansion of what had hitherto been classified as feeder ports. These became potential main-line ports of call and/or hub ports in a secondary network containing feeder services to a fast-expanding third tier of feeder ports in the region.

Whilst this analysis of Robinson (1998) did include some consideration of the Chinese mainland container ports, the main focus was very much on the whole of the Asian region *in toto*. As such, his conceptualization of the development of container ports may not be wholly applicable to the very

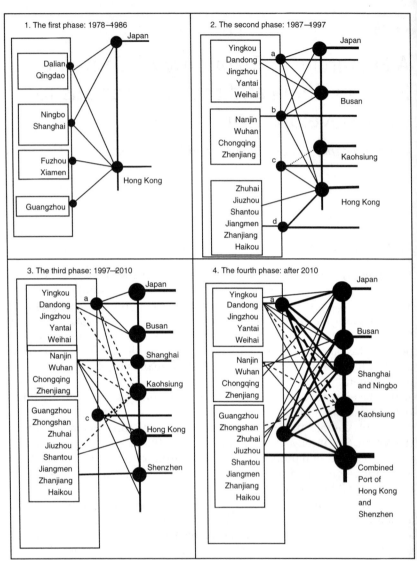

The width of the line represents the volume of containers on this route
A weak link between two ports (Point-to-point shipping)
Possible future direct links and current indirect shipping routes.
a: Dalian, Qingdao, Tianjin; b: Ningbo, Shanghai; c: Fuzhou, Xiamen;
d: Shenzhen, Guangzhou, Zhongshan

Figure 13.1   Phases in the development of container ports in China

special situation of China. A similar approach, however, can be found in the China Shipping Development Annual Report (Department of Water Transport, 1998) and is based on dividing the development of China's ports into three phases. Because the analysis presented in the report does not focus on the development of dedicated container ports or terminals, but deals instead with the general development of China's ports, irrespective of their particular specialization, it similarly does not provide a wholly appropriate analysis of China's container ports sector.

The development of container ports in China is divided into four distinct phases in line with the conceptual framework presented in Figure 13.1; one which is itself based closely on the conceptualization due to Robinson (1998).

This chapter goes on to identify the way that port policies have influenced the development of China's major ports and the shuttle lines which serve them. Finally, a hierarchy of ports is developed. This latter is especially important since it is this which exerts a major influence over cargo flows and thus the whole structure of the liner shipping network which services China's trade.

13.2   Phases in the development of China's container ports

Phase 1: 1978–1986

In 1978, China's state-owned shipping company (China Ocean Shipping Company – COSCO) inaugurated China's first venture into the container transport business with a maiden voyage from Shanghai to Australia (Department of Water Transport, 1998). This marked the true start of container port development in China despite the fact that over the whole of this phase, from 1978 to 1986, almost no *dedicated* container terminals were established. Instead, during this period an emphasis was placed on the use of bulk or general cargo quays for the berthing of container ships.

The capacity and quality of service of these ad hoc 'container ports' could not meet the burgeoning demand for the transport of containerized cargoes. The problem became acute when serious port congestion occurred in 1981, 1983 and 1985. The inadequacy of container handling capabilities in ports had effectively become a major bottleneck, thus restricting foreign trade and the further economic development of the Chinese mainland. Despite these problems and albeit from a low base, the growth rate in container throughput over this period, as shown in Figure 13.2, was rather impressive. Perhaps, however, the adverse impact of poor container handling facilities can be witnessed in the dampened growth rate that occurred towards the end of this phase.

The growing importance of containerized cargoes to the ports sector, shown in Figure 13.3, reflects the degree of acceptance that the containerized transport of international cargoes was slowly but surely to achieve in

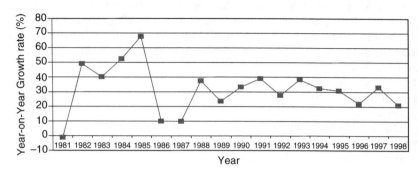

Figure 13.2 Growth rates of container throughput, 1981–1998
Source: Department of Water Transport (1998).

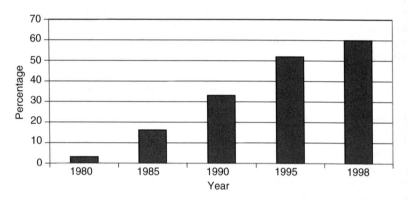

Figure 13.3 Importance of containerized cargoes to China's main ports, 1980–1998 (percentage of total tonnes handled)
Source: Department of Water Transport (1998).

the Chinese mainland over this period and later. According to Yeung (1996), however, since there were no real regional container ports or international liner services, during this phase most containers were shipped to either Hong Kong or Japan for transhipment onto mainline services.

Phase 2: 1987–1997

As early as 1985, China's state council promulgated the provisional rules for the 'Favourable Treatment for Construction of Terminals by Sino-Foreign Joint Ventures'.

Under this set of rules, the construction of terminals by joint ventures involving Chinese mainland and overseas collaboration was not only permitted but positively encouraged. In effect, this set of rules provided a policy guarantee which underpinned the rapid development of container

Table 13.1 Developments of dedicated container terminals and container throughput in the Chinese mainland, 1987–1997

Year	Number of dedicated terminals	Designed container throughput (thousand TEU)	Actual throughput (thousand TEU)	Growth rate (%)	Proportion of containerized cargoes in coastal ports (%)
1987	14	1,100	689	9.89	20.0
1988	15	1,200	947	37.45	23.0
1989	18	1,330	1,170	23.55	26.0
1990	19	1,450	1,560	33.33	33.0
1991	23	1,950	2,170	39.10	42.0
1992	34	3,120	2,770	27.65	43.0
1993	36	3,540	3,830	38.27	44.0
1994	42	4,340	5,070	32.38	46.0
1995	52	5,330	6,630	30.77	52.0
1996	57	6,380	8,090	22.02	57.0
1997	65	10,030	10,770	33.13	n.a.

Source: Department of Water Transport (1998).

ports in the Chinese mainland. Table 13.1 shows the development of dedicated container terminal and container throughput in the Chinese mainland 1987–1997.

The beginning of the second phase in the development of container ports in the Chinese mainland began in earnest in 1987 with the establishment of the Nanjing International Container Terminal Company Ltd. This was the first Sino–foreign container handling enterprise on the Chinese mainland and involved a joint venture between Nanjing Port Authority and US-based Encinal Terminals. The main characteristics of this phase in container port development can be described as follows:

- *The construction of new container ports.* During this phase of development, there was a good match between the designed capacity of the main container ports and actual container throughput. Even though from 1993 overall designed capacity lagged behind actual throughput, this problem of insufficient capacity had been largely rectified by 1997. Clearly, during this phase, much attention was paid to the planning and the construction of dedicated container terminals to meet the needs of the Chinese mainland's international trade. A comparison of designed and actual container throughput in the main container ports can be found in Figure 13. 4.
- *Privatization and commercialization of China's container ports.* Since 1985, China has invested more in its port development than the rest of the world combined (Frankel, 1998). To attract more capital, China has moved

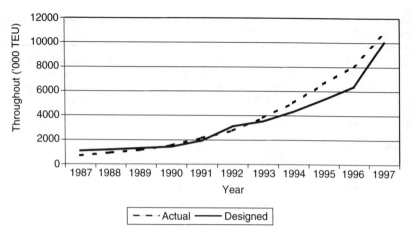

Figure 13.4   A comparison of designed and actual container throughput
Source: Department of Water Transport (1998).

rapidly towards the privatization of its ports and, particularly, its container terminals. In fact, over the last twenty years, China's port system has experienced a shift from a policy of centralized control to one, which is much more decentralized (Department of Water Transport, 1998). In addition, since the setting up of the first Sino–foreign International Container Terminal Company Ltd in Nanjing in 1987, an increasing amount of overseas capital has been invested in the development of container terminals (Frankel, 1998).

• *Upgrading of certain feeder ports to regional hub ports.* Prevailing market conditions and the vigorous competition between liner shipping companies during this period has meant that the container shipping industry has emerged as only a marginally profitable business. As a result, carriers have focused their energies on pursuing market share through cost-cutting. In addition, mergers, takeovers and alliances amongst the larger liner shipping organizations have consolidated the market domination of a few large companies (Ryoo and Thanopoulou, 1999). These alliances have redeployed their fleets and reconfigured and rescheduled their services and, by so doing, have led to a worldwide rationalization of container transport so that fewer and fewer container ports are called at directly by mainline vessels (Cullinane and Khanna, 1999).

   In stark contrast to the hub concentration which has resulted from this worldwide process of fleet and schedule rationalization, certain container feeder ports in China (for example, Shanghai and Shenzhen) have gradually emerged as regional hub ports. In part, this reflects China's increasing market orientation, its adoption of economic liberalization policies and the ending of its economic and political isolation. From a more pragmatic and transport-oriented perspective, however, it also reflects its emergence as the world's manufacturing powerhouse and, concomitant with this, as the world's largest potential consumer market.

- *The development of China–Korea and China–Taiwan shipping services.* The Chinese and South Korean governments established diplomatic relations in 1992. At that time, some large container ports in South Korea (but most notably Busan) had become international hub ports (Cullinane and Song, 1998). The increasing trade between China and South Korea, with Busan container port acting as the international transhipment centre, greatly stimulated the development of China's container ports, especially Qingdao, Tianjin, Dalian and Shanghai, as well as many medium and small ports in Shangdong and Liaoning provinces.

In 1997, following protracted negotiation between the governments of both China and Taiwan, experimental direct sailing was introduced across the Taiwan Strait by ten Chinese and Taiwanese shipping companies. The potential for container transport between China and Taiwan is extremely promising. This is not only because of Kaohsiung's existing position as a major international hub port but also because of the forecast growth of Xiamen and Fuzhou container ports which lie across the Taiwan strait in the Chinese mainland.

Phase 3: 1997–2010

Two policies promulgated by the Ministry of Communications in 1997 marked the beginning of the third phase. The first of these was the imposition of cabotage restrictions, which reserved market entry onto coastal shipping routes solely for vessels flying the flag of the Chinese mainland. The avowed intention of this policy was to encourage the initiation of coastal feeders and, thereby, to provide support to the development of major ocean liner routes. The second policy, which might appear to contradict the first, was to impose a generally applicable 20 per cent increase in port charges for vessels engaged in coastal shipping services. This could be somewhat cynically viewed, however, as the price which local operators had to pay for the protection from overseas competition, which the first of these policies rendered them.

During this phase, and taking many years for their construction, several large container terminals have been established in the Chinese mainland. Plans for their further development to 2010 are already in place and construction activities, especially with respect to the provision of appropriate transport infrastructure to serve these container terminals, also have a long planning horizon. Figure 13.1 has already shown the large coastal ports and main liner services extant during this

Table 13.2 Throughput of the Chinese mainland's top 10 container ports ('000 TEU)

Port	1981	1985	1990	1995	1996	1997	1998	1999	2000	Growth rate (compared with 1999) (%)
China	104*	503*	1,560	6,630	8,090	10,765	13,124	18,059	23,480	30
Shanghai	49	202	456	1,527	1,971	2,527	3,066	4,216	5,612	33
Shenzhen	0	0	33	284	589	1,147	1,952	2,986	3,994	34
Qingdao	12	33	135	603	810	1,033	1,213	1,542	2,120	37
Tianjin	26	190	286	702	823	936	1,018	1,302	1,708	31
Guangzhou	11	47	110	515	558	687	841	1,177	1,431	22
Xiamen	0	0	36	310	400	546	654	848	1,085	28
Dalian	6	30	131	374	421	453	526	736	1,011	37
Ningbo	0	1	22	160	202	257	353	601	902	50
Fuzhou	0	0	29	151	177	225	252	318	400	26
Zhuhai	0	0	0	275	270	264	262	291	314	8

Source: Department of Water Transport (2000).
*Estimated by the authors.

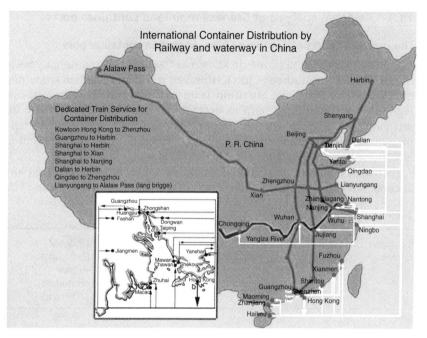

Figure 13.5 The three major geographical port groupings in mainland China

period (phase 3) and alludes to the future expectations, which are embodied in the currently planned infrastructure investment (phase 4). Table 13.2 shows the throughput of the top ten container ports from 1981 to 2000 and, in so doing, highlights the progress which is being made, not only in economic development but also in meeting the logistics needs for facilitating this development.

According to Table 13.2, in 2000 the market share (of the Chinese mainland's total throughput) of the top five and top ten container ports accounts for 63 per cent and 79 per cent respectively. In effect, three major regional groupings of container ports have taken shape in the Chinese mainland over this period according to their geographic location and container throughput at ports. They are:

- Southern China: Shenzhen, Guangzhou, Zhongshan.
- Central China: Shanghai, Ningbo.
- Northern China: Qingdao, Tianjin and Dalian.

The geography of these three important container port groupings within the Chinese mainland is depicted in Figure 13.5.

13.3   Regional analysis of Chinese mainland container ports

The container ports of Southern China; Shenzhen container port

The port of Guangzhou (formerly known as Canton) has traditionally been the largest port in southern China. However, it is now Shenzhen container port which attracts most attention from industry commentators in the region. This is because of both its exceptional rate of growth and also its intriguing relationship with the port of Hong Kong in terms of an overlapping ownership structure and its position as Hong Kong's main competitor and major potential collaborator (Cullinane, 2000).

Shenzhen container port comprises the container terminals of Yantian, Shekou, Kaifeng and Chiwan, the details of which are shown in Table 13.3.

Table 13.3   Container terminals in Shenzhen port

Container terminal	Container berths	New or under construction berths	Designed berthing capacity (1000 tons)	Designed throughput volume per annum (1000 teu)
SCT	2	0	50	500
Kai Feng	2	1	50	500
Chi Wan	0.5	0	75	150
Yan Tian	2	3	50	500
Shekou	0	3	n. a.	n. a.
Ma Wan	0	0	n. a.	n. a.
Da Pen Bay	0	0	n. a.	n. a.
Total	6.5	7		1,650

Source: Department of Water Transport (2000).

Table 13.4   Liner services calling at Shenzhen container port as of October 1999

Route	Frequency	Liner	Ship slots
America	Once a week	Global Alliance	4,300
	Once a week		4,832
	Once a week	Grant Alliance	3,600
	Once a week		4,830
	Once a week	K-line/Cosco/Yangming	3,720
	Once a week	Evergreen	5,364
	Once a week	Maersk-Sealand	6,000
	Once a week		2,408
	Once a week		6,600
	Once a week		1,800
	Once a week	Pan Pacific	1,728
	Once a week	North Europe-Asia/ CMA(CGM)	900

Table 13.4   Liner services calling at Shenzhen container port as of October 1999 – *continued*

Route	Frequency	Liner	Ship slots
	Once a week	Mediterranean	2,206
	Once a week	HKSQ	900
	Once a week	Cosco	5,250
	Once a week	Zim	3,416
	twice a week	Pacific	1,986
Europe	Once a week	Maersk-Sealand	4,300
	Once a week	Grant Alliance	4,600
	Once a week		4,200
	Once a week	Global Alliance	3,980
	Once a week	Mediterranean	4,000
	Once a week	North Europe-Asia/ CMA(CGM)	4,000
	Once a week	China Shipping Co.	2,097
Mediterranean	Once a week	Zim	3,016
Southeast Asia	Once a week	Grant Alliance	4,960
	Once a week	Global Alliance	4,481
	Once a week	Pacific	1,650
	Once a week	Zim	670
	Once a week		568
	Once a week	Cosco	1,200
	Once a week	China Shipping Co.	1,000
Australia	Once a week	Russo-Orient	1,748
	Twice a week	ANL	2,825

Source: Department of Water Transport (1999).

Table 13.5   Forecast changes in the hinterland of Shenzhen container port

1995 Hinterland	%*	2010 Hinterland	%*
Shenzhen city	60%	Shenzhen city	42%
Pearl River delta (except Shenzhen); Coastal cities and regions along Guangdong province; Regions along east–north coast; Guangxi autonomous region Jiangxi province; Regions along Beijing– Guangzhou, etc	40%	Most regions in Guangdong province; Regions along the railway of Beijing–Guangzhou, Beijing–Kowloon; Regions along southeast coast Hong Kong and some East-Asian areas.	58%

*The percentage share of the total containers moving through the container ports of Shenzhen that are generated in a region

Generally speaking, the terminals of Shenzhen container port used to fulfil a feeder function for the port of Hong Kong, as shown in Figure 13.1. This situation, however, is changing rapidly as they develop still further. As argued by Slack (1998), Hong Kong's pre-eminent position as China's most important hub port is due to the competitive weakness of other ports, rather than to any inherent characteristics of containerization. The hinterland of Shenzhen container port will expand with its continued growth and eventual maturity. As at the time of writing, there are altogether 156 container ships calling at Shenzhen container port, including 73 which are over 4,000 TEUs, 61 of 3,000–4,000 TEUs and 22 less than 3,000 TEUs. Table 13.4 illustrates the international liner services calling at Shenzhen container port.

An official forecast of the development of the future hinterland of Shenzhen container port (prepared by Wang, 1998) is presented in Table 13.5.

Coastal container ports in southern China

The Pearl River Delta (PRD), which includes Guangdong and Hainan Provinces and the Guangxi Zhuang Autonomous Region, has always been regarded as the area of the Chinese mainland which is at the forefront of implementing China's policies for economic reform. In 1998, 60 per cent of the Chinese mainland's total container throughput either originated from, or was destined for, this area. Following Hong Kong's return to China on 1 July 1997, the relocation of Hong Kong's manufacturing industry across the border to the PRD area has accelerated and this has unquestionably stimulated a significant growth in demand for container transportation and, thus, the development of the region's ports and container terminals.

Table 13.6   Throughput of the main south China container ports, 1995–2000 (TEUs)

Port	Year					
	1995	*1996*	*1997*	*1998*	*1999*	*2000*
Shenzhen	283,681	589,057	1,147,347	1,951,746	2,824,000	3,994,000
Guangzhou	514,987	557,528	687,303	840,000	1,120,000	1,431,000
Zhongshan	178,176	221,781	315,530	378,636	415,000	506,000
Zhuhai	274,637	270,095	261,985	257,570	291,000	314,000
Shantou	69,742	90,016	74,228	63,223	117,000	114,000
Jiangmen	36,622	33,735	42,349	50,819	n. a.	n. a.
Zhanjiang	29,944	29,465	43,672	33,089	49,000	75,000
Huicheng	16,363	25,093	27,832	n. a.	n. a.	n. a.
Haikou	20,702	16,637	26,047	28,694	38,000	49,000
Total	1,470,574	2,074,151	2,781,281	3,603,777	n. a.	n. a.

Source: Department of Water Transport (1998, 2000).

Since 1990, there has been a rapid increase in the volume of containers moved to and from the provinces of Guangdong and Hainan and to and from the Guangxi Zhuang Autonomous Region. This new volume has been fed into the container terminals of Shenzhen and the rest of southern China, largely via coastal feeder services. Meanwhile, Yunnan, Guizhou and Sichuan provinces and the southern region of the Yangtze River are fast becoming new economic hinterlands for these ports. Table 13.6 shows the development of the container ports in southern China.

Ports along the Pearl River

The Pearl River mainly comprises Xijiang, Beijiang and Dongjiang. It flows across Guangdong province and the Guangxi Zhuang Autonomous Region and constitutes the main inland waterway in southern China. There are altogether 57 ports along the river, most of which are extremely small (Wong, 1996).

Containerized river transport between these small ports (including Shekou, Ma Wan, Chi Wan and Zhuhai) comprises a fleet of river boats ranging from 10- to 200-box capacity. This transport mainly acts as a feeder service for the regional large container ports. In fact, according to Tai-Yuan and Beresford (1995) the Pearl River has not been fully exploited in terms of inland water transportation.

The container ports of central China: Shanghai container port

As mainland China's traditional industrial centre, Shanghai is often the focus of worldwide attention. As shown in Table 13.7, the throughput of Shanghai container port has developed dramatically since the reform of the Chinese economy.

Table 13.7   Throughput of the Shanghai container terminals, 1980–98 ('000 TEUs)

Year	Imports	Exports	Throughput	Growth rate (%)
1980	15.9	14.5	30.4	—
1985	106.9	94.8	201.7	46.0
1990	224.3	231.8	456.1	17.7
1991	281.3	295.4	576.7	26.4
1992	339.4	391.1	730.5	26.7
1993	448.3	486.4	934.8	28.0
1994	555.7	643.5	1,199.2	28.3
1995	693.0	833.5	1,526.5	27.3
1996	924.1	1,047.3	1,971.4	29.1
1997	1,146.6	1,380.7	2,527.3	28.2
1998	1,411.0	1,655.0	3,066.0	21.3

Source: Guo (1999).

Table 13.8   Container terminals in the port of Shanghai

Terminal	Berths	Length (metres)	Designed capacity (TEU)	Throughput in 1998 (TEU)
Zhang Hua Bang Terminal	3	784	800,000	910,000
Jun Gong Lu terminal	4	857	650,000	844,000
Baoshan terminal	3	640	250,000	273,000
Wai Gaoqiao (first phase)	3	900	600,000	675,000
Others				358,000
Total	13			3,060,000

Source: Guo (1999).

Table 13.9   Liner services calling at Shanghai container port as of the end of December 1999

Route	Frequency	Liner	Ship slots
America	Once a week	Cosco	1,000
(west coast)	Once a week	Yangming	1,000
	Once a week	Maersk/Sea-land	1,000
	Once a week	CMA(CGM)	600
	Once a week	Mediterranean	600
	Once a week	NYK	900
	Once a week	MOSK	1,000
	Once a week	P&O N	900
	Once a week	Zim	300
	Once a week	Hanjin	1,200
America	Once a week	Cosco	1,000
(East coast)	Once every two weeks	HKSQ	1,100
Hong Kong	Once a week	Ever Green	600
	Once a week	Haihua	288
	Once a week	Xinhai	327
	Once a week	HK Orient Transportation	380
Europe	Once a week	American President	1,000
	Once a week	Hapag-Lloyd	1,800
	Once a week	CMA(CGM)	400
	Once a week	Hyundai	1,000
	Once a week	MOSK	1,000
	Once a week	OOCL	1,800
	Once a week	Mediterranean	600
	Once a week	Cosco/K-line	1,050
	Once every two weeks	Lloyd Triestino	600
	Once every two weeks	China Shipping Co.	700
	Once a month	Hanjin/Sino-Trans	700
	Once a month	Maersk/Sealand	1,000

Table 13.9   Liner services calling at Shanghai container port as of the end of December 1999 – *continued*

Route	Frequency	Liner	Ship slots
Mediterranean	Once a week	Zim	800
	Once a week	Cosco	670
	China Shipping Co.	400	
Southeast Asia	Twice a week	Global Alliance	400
Australia	Once a week	Cosco	350
	Once a week	China Shipping Co.	200
	Twice a week	ANL	300
Japan	Once a week	Cosco	400
	Once a week	China–Japan International Ferry Co.	224
	Once a week	China Shipping Co.	200
	Once a week	Shanghai International Ferry Co.	229
	Once a week	Xinhai	480
	Once a week	Tianhai	617
	Once a week	K-line	450
	Twice a week	Central Asian Shipping	20
	Twice a week	Jinjiang	443
	Twice a week	Yantai Shipping Co.	310
	Four times a week	Sino-trans	310
	Once every ten days	Haihua	552
	Five times a week	Minsheng Kambara Marine Shipping.	150
Korea	Once a week	Korea Marine Transport Co.	500
	Once a week	China Shipping Co.	200
	Once a week	Jinjiang	300
	Once a week	Xindong	762
	Once a week	Cosco	672
	Once a week	Changjin	380
	Twice a week	Dongying	475
East and west of Africa	Once every 10 days	MOSK	88
	Once a week	P&O N	65
	Once a week	Delmas	50

Source: Department of Water Transport (1999).

In 1998, the port of Shanghai included 13 dedicated container berths distributed within Baoshan Terminal, Jun Gong Lu Terminal, Zhang Hua Bang Terminal and Wai Gaoqiao terminal. The total designed annual throughput capacity was 2.3 million TEUs and encompassed the deployment of 22 gantry cranes. The details are shown in Table 13.8.

However, a maximum draught restriction of 11 metres greatly impedes the further development of Shanghai container terminal as a truly international container hub port. Table 13.9 shows the details of container ships

Table 13.10   Committed and planned container port projects in, Shanghai

Project	Quay length (metres)	Annual capacity (m TEUs/year)	Completion by end
Waigaoqiao: 5 spp QCCs		0.5	2000
4 spp QCCs		0.4	2001
Phase III	665	0.6	2001
Phase IV	665	0.6	2002
Wahaogou: new terminal	600	0.4	2003
	600	0.4	2003
Jinshanzui: new terminal	600	0.4	2006
	600	0.4	2008
Xiaoyangshan/Dayangshan	2,000	2.5	2007–10
		12.5	After 2010

Source: OSCL (2001: 80).

calling at Shanghai container terminal. According to this table, in sharp contrast with Shenzhen container port which is called at by most large container ships, it is clear that the largest ship only has a slot capacity of about 1,800 TEUs.

Table 13.10 shows the new projects underway or being planned in the Yangtze River estuary to solve this problem. It is of great significance to note that a new purpose-built deep-water container terminal with a maximum draught restriction of 15 metres in Yang Shan, as shown in Table 13.10, is under construction. This new project, scheduled to be used after 2010, will fundamentally overcome the drawback with Shanghai container port and restructure the worldwide container transportation network. This restructuring is also illustrated in Figure 13.1.

Ningbo container port

Beilun container terminal constitutes the main part of Ningbo container port. It comprises two dedicated container berths with designed annual throughput capacity of 0.5 million TEUs. The water depth is 13.5 metres, which means that a Panamax containership of 80,000 deadweight tons (dwt) is the maximum size of vessel which is capable of berthing here.

Because of the comparatively short distance between the container ports of Shanghai and Ningbo (204 km by rail), their close cooperation is encouraged by the Chinese government. In September 1997, a cross-regional container terminal administrative organization, the Shanghai Port Group, was established to regulate competition and to maintain and promote the pace of development of container ports in Shanghai, Zhejiang and Jiangsu provinces. However, because China's current port policies are characterized by a decentralized approach, which stimulates competition, the management of ports and terminals have appeared to pay much more attention to establishing themselves as regional hub ports. Hence, the feasibility of this form of cooperation proving successful is rather dubious.

Container transport along the Yangtze River

The Yangtze is the longest river in China. It flows for 2,815 km and, altogether, there are 26 ports along it, including some relatively large ports such as Nanjing. In 1998, the containers transported along this East–West corridor accounted for 80 per cent of total containers transported by inland waterway. The container throughputs of the main ports along the Yangtze River are shown in Table 13.11.

The main shipping companies involved in moving cargoes through the Yangtze River are the China Changjiang National Shipping Corporation (China's largest inland shipping group) and the China Ocean Shipping Company (COSCO – the national carrier) with over 40 barges deployed in an extensive network of barge/river ship services.

Fossey (1998) highlights the importance of the Yangtze River by pointing out that the government intends it to become the intermodal corridor which links such inland cities as Chongqing, Changsha, Wuhan and Wuhu. Following the completion of the much-heralded, controversial and extremely expensive 'Three Gorges Dam' project, the flow of the river will be much more controllable and average water depth up to the city of Chongqing (the largest city in China with more than 15 million people) will be raised by approximately three metres. This will allow barges and coastal ships up to a maximum of 10,000 dwt to sail from the seaports to

Table 13.11   Container throughputs of the main ports on the Yangtze River, 1986–2000 (TEUs)

Year	Nantong	Zhangjiagang	Nanjing	Wuhan
1986	5,188	33,034	379	238
1987	11,451	35,891	1,016	1,370
1988	17,271	38,786	4,525	3,263
1989	9,435	47,985	23,946	2,552
1990	10,907	49,990	42,021	1,605
1991	19,999	60,375	51,797	2,548
1992	30,046	67,017	73,303	6,205
1993	46,534	81,964	109,098	14,120
1994	65,899	94,587	126,213	16,343
1995	87,179	108,063	144,657	13,766
1996	93,082	118,224	130,287	16,932
1997	71,591	119,132	130,266	18,568
1998	120,362	105,051	123,218	18,659
1999	158,000	113,000	157,000	25,000
2000	182,000	137,000	203,000	30,000

Source: Department of Water Transport (2000).

Chongqing for the first time. This is the keystone element in the Chinese government's long-term objective to industrialize and open up the country's interior, while simultaneously transforming Shanghai into an international shipping centre.

Because of the importance of the Yangtze River corridor and China's obvious determination to develop it as one of its main economic arteries, ports and other facilities along the Yangtze River have attracted much investment and their number and sophistication has increased accordingly.

The container ports of Northern China: Qingdao, Tianjin and Dalian

The container throughput of Qingdao is greater than that of Tianjin, Dalian and other northern container ports. This is mainly due to: the rapid development of its hinterland economy, especially within the city of Qingdao itself; its natural geographically advantageous situation; the 14.5-metre water depth in Qingdao container port can be utilized for berthing 5th generation containerships over 5,250 TEUs (Chadzynski, 1997); the fact that port users regard the management of the port as highly efficient and effective (Anon, 1997).

The container throughputs of Tianjin and Dalian container ports rank second and third respectively in northern China. The designed throughput capacity in Tianjin is 1.4 million TEUs and the water depth is 12~14 metres. Dalian Container Terminal is a joint venture between the Port of Singapore Authority (PSA) and Dalian port authority. Its designed annual throughput capacity is 1.15 million TEUs and with a water depth of 12.1–14 metres, it is capable of berthing 5th generation containerships.

Despite highly efficient management in both Tianjin and Dalian container ports, the economic development of their most proximate and main hinterlands has progressed only comparatively slowly in recent years and it is difficult to envisage how it can improve further. This is especially the case for Liaoning province, formerly one of the centres of heavy industry on the Chinese mainland. Aside from the major portions of each of their hinterlands, these three container ports also access cargoes sourced from or bound for some regions where the hinterlands of the three ports are overlapping. This means that there will be vigorous competition between the three of them, at least for some time into the near future.

Medium and small container ports in Northern China

There are many medium or small ports such as Yingkou and Yantai around Bohai Bay and along the coast of the Yellow Sea. In terms of container transport, they mainly serve as feeder ports for Qingdao, Dalian and Tianjin. They also possess some direct services to South Korea and Japan, although their market share is fairly small (Anon, 1999).

13.4 Discussion

A continuation of the Chinese mainland's policies of economic and political liberalization will mean that the same trends which are now impacting upon the global ports sector will inevitably emerge within the context of the Chinese ports sector.

Indeed, the foregoing analysis suggests that the early stages of some of these trends have already appeared. In particular, having recognized the benefits of private sector participation in the ports and container terminals industry, the Chinese authorities have stimulated what is now a significant financial and managerial commitment to the sector from both China's own commercial private sector enterprises and from global port owner-operators such as Hutchinson Port Holdings, PSA, P&O Ports and Modern Terminals Ltd. The attraction of foreign finance has even extended to international logistics companies, such as Maersk and Kerry, who have also involved themselves in either or both of the ownership or management of Chinese mainland container ports and terminals (Song, 2001).

Closely associated with the increased participation of private sector interests in the container ports of the Chinese mainland, there is also a burgeoning demand for, and provision of, *dedicated* container terminals. This also reflects a trend which is taking place worldwide.

It will be extremely interesting to see whether the explosion in China's international trade and the greater liberalization of its trading environment, that are both anticipated to follow China's accession to the World Trade Organization, will provide a simultaneous boost to foreign direct investment in the Chinese mainland's ports sector. Potential investors will surely be attracted by the prospect of booming trade in an economically and politically more liberal and secure business environment where both bureaucracy and inefficiency have been greatly reduced.

Although covering a very large and geographically diverse area, it has already become apparent that there is much greater competition at the regional level than was previously the case in the Chinese mainland's ports sector. The further inculcation of competition into the ports sector, which overseas private sector participation will bring, is likely to intensify competition still further. Initially, this is likely to be most acutely felt within each of the three major regional clusters of port activity. However, as time progresses and as extant plans for transport infrastructure improvements come to fruition, this will lead inevitably to a situation where competition within China's ports and terminals sector becomes trans-national in scope.

The resolution of the competitive forces at play in the marketplace will be greatly complicated by the fact that many of the new private sector actors involved in the Chinese market have investments in numerous different ports and terminals, some of which are in direct competition with each other. It will be complicated still further if the growing worldwide tendency towards the strategic alliance of container ports and terminals is adopted and applied within China. Were such a situation of *co-opetition* (Jorde and Teece, 1989; Brandenburger and Nalebuff, 1996) to emerge within China and elsewhere, it implies a borderless port community with little or no national responsibility and accountability; a phenomenon which starkly contradicts the traditional view of ports as the gateways for a nation's trade and of ports policies as instruments for maximizing national welfare.

Within the wider context of China as a whole, it is the current situation of Hong Kong which is particularly intriguing. Focusing on China's container ports sector from a purely national perspective, it is clearly the case that Hong Kong's cost-competitiveness is severely undermined by any comparison to the container ports and terminals of the Chinese mainland, especially those within its own hinterland of Shenzhen (Cullinane, 2000). At the same time, the advantage that Hong Kong possesses in terms of service quality is being rapidly eroded. What may ultimately prove to be the saviour of Hong Kong is the role which it plays in allowing China to compete internationally in the ports sector. At least in the short to medium term, its geographical characteristics, including its physical location, are such that it is has major advantages in seeking to maintain its international status as a major hub port for Asia (see Cullinane and Khanna, 2000).

Attempts by other ports in the Chinese mainland (but particularly those located in the central and northern port clusters) to establish themselves as major hub ports for Asia or, eventually, even to maintain their positions as regional hubs for China's trade, are likely to be severely undermined by the enormous and unyielding competition which they will face from Kaohsiung and Busan.

If the Chinese mainland ports are to compete internationally, efficiency must continue to improve by leaps and bounds. In line with the economic theories of public choice and property rights (Hart and Holmstrom, 1987; Shapiro and Willig, 1990; Martin and Parker, 1997), it is to be expected that greater private sector participation will improve productivity levels within China's ports. However, there is also a very clear-cut relationship between the scale of operation and productivity (Tabernacle, 1995) and it may well be the case that, with the exception of Hong Kong, China's container ports have left it rather too late to mount a concerted effort to compete in the international market for hub port status. In this context, the market domination of Hong Kong, Kaohsiung and Busan are already too ingrained. In particular, should some political reconciliation be reached between Taiwan and the Chinese mainland, simply the geographical position of Kaohsiung relative to the major internal trade flows of China (Cullinane, Ji and Wang, 2002) will place it in an enviable and dominant market position.

13.5   Conclusions

On the basis of existing policy on the Chinese mainland, this chapter has tentatively proposed a third phase of container port development to continue until 2010. It remains to be seen whether the conditions negotiated for China to become a member of the WTO will bring about such a radical change in China's ports policies that a new phase in container terminal development will be the result.

Irrespective of WTO membership, as things stand and if current trends continue, there is no doubt that container ports and terminals in the Chinese mainland will benefit still further from the injection of overseas investment and expertise. There remains plenty of scope for this to occur. As a result, one can expect to see productivity levels continue to increase and the container operations in ports to become more seamlessly integrated with an ever-improving land-based freight transportation infrastructure.

There are a number of potential influences which might deflect this extrapolation of the development of China's container ports and terminals, all of which provide a fertile ground for further research. With the focus being the broad outcome of the competition between the different container ports within China, the ultimate aim must be to predict the market share of individual ports. On the basis of the arguments presented herein, it is clear that this is not going to be an easy task. A number of potential influences have been identified which, although capable of being analysed independently, are also interrelated and, therefore, have a combined effect. In particular, the individual and combined impact of WTO membership on the volume and nature of trade and industrial location; improvements to the transport infrastructure and the level of cooperation between ports would also seem to be areas where further work would prove especially beneficial.

Note

1   This chapter includes material drawn from Cullinane, Wang and Cullinane (2004). The editors are grateful to Taylor & Francis for granting permission to reproduce it herein.

References

Anon (1997) Qingdao's ship comes in, *China Daily*, 19 October.

Anon (1999) Liner Shipping Routes, *Containerisation International Yearbook*, London, p. 264.

Brandenburger, A.M. and Nalebuff, B.J. (1996) *Co-opetition*, New York: Currency.

Chadzynski, W. (1997) Container Vessels 1990–1996: Design, Characteristics, Trends, Development, in T. Graczyk, T. Jastzebski, and C.A. Brebbia (eds), *Marine Technology II*, Southampton: Computational Mechanics Publications.

Cullinane, K.P.B. (2000) The Competitive Position of the Port of Hong Kong, *The Journal of the Korean Association of Shipping Studies*, 31, 45–61.

Cullinane, K.P.B. and Khanna, M. (1999) Economies of Scale in Large Container Ships, *Journal of Transport Economics and Policy*, 33(2), 185–208.

Cullinane, K.P.B. and Khanna, M. (2000) Economies of Scale in Large Containerships: Optimal Size and Geographical Implications, *Journal of Transport Geography*, 8, 181–95.

Cullinane, K.P.B. and Song, D-W. (1998) Container Terminals in South Korea: Problems and Panaceas, *Maritime Policy and Management*, 25(1), 63–80.

Cullinane, K.P.B., Ji, P. and Wang, T. (2002) A Multi-Objective Programming Approach to the Optimisation of China's International Container Transport Network, *International Journal of Transport Economics*, 29(2), 181–99.

Cullinane, K.P.B., Wang, T. and Cullinane, S.L. (2004) Container Terminal Development in Mainland China and its Impact on the Competitiveness of the Port of Hong Kong, *Transport Reviews*, 24(1), 33–56.

Department of Water Transport (1998, 1999, 2000) *China Shipping Development Annual Report*, Beijing: Ministry of Communications, People's Republic of China.

Fossey, J. (1998) China's Ports Surge Ahead, *Containerization International*, July.

Frankel, E.G. (1998) China's Maritime Developments, *Maritime Policy and Management*, 25(3), 235–49.

Guo, X.L. (1999) Analysis of the Feasibility and Urgency of Building Shanghai Port into a Deep-water Port based on its Current Situation, *China Ports Special*, 38–9 (in Chinese).

Hart, O.D. and Holmstrom, B. (1987) The Theory of Contracts, in T.F. Bewley (ed.), *Advances in Economic Theory*, Cambridge: Cambridge University Press.

Jorde, T.M. and Teece, D.J. (1989) Competition and Co-operation: Striking the Right Balance, *California Management Review*, 31(3), 25–37.

Martin, S. and Parker, D. (1997) *The Impact of Privatisation: Ownership and Corporate Performance in the UK*, London: Routledge.

OSCL (2001) *The East Asian Containerport Market to 2015*, Chertsey: Ocean Shipping Consultants Ltd.

Robinson, R. (1998) Asian hub/feeder nets: the dynamics of restructuring, *Maritime Policy and Management*, 25(1), 21–40.

Ryoo, D.K. and Thanopoulou, H.A. (1999) Liner Alliances in the Globalization Era: a Strategic Tool for Asian Container Carriers, *Maritime Policy and Management*, 26(4), 349–367.

Shapiro, C. and Willig, R.D. (1990) Economic Rationales for the Scope of Privatization, in E.N. Suleiman and J. Waterbury (eds), *The Political Economy of Public Sector Reform and Privatization*, Boulder: Westview Press.

Slack, B. (1998) Intermodal Transportation, in B. Hoyle and R. Knowles (eds) *Modern Transport Geography*, Chichester and New York: Wiley.

Song, D.-W. (2001) Regional Container Competition and Co-operation: the Case of Hong Kong and South China, *Journal of Transport Geography*, 10(2), 99–110.

Tabernacle, J.B. (1995) Changes in Performance of Quayside Container Cranes, *Maritime Policy and Management*, 22(2), 115–24.

Tai-Yuan, J. and Beresford, A.K.C (1995) *Freight Transport in China, An Overview*, Dept. of Maritime Studies and International Transport, College of Cardiff, University of Wales.

Wang, W.G. (1998) Study on Developing Shenzhen into an International Shipping Centre, *Proceedings of the International Conference on World Shipping and Shipping Market Facing 21st Century*, Shenzhen, China (in Chinese).

Wong, B.P. (1996) Hong Kong's Port in the Pearl River Delta Region, in Y.-O. Geh (ed.), *Planning Hong Kong for the 21st Century*, Hong Kong: Hong Kong University Press.

Yeung, Y.-M. (1996) Hong Kong's Hub Functions, in Y.O. Geh (ed.), *Planning Hong Kong for the 21st Century*, Hong Kong: Hong Kong University Press.

Index